Molecular Developmental Biology

The Forty-Third Symposium of
The Society for Developmental Biology
Columbia University, June 18–20, 1984

Executive Committee
1983–84

1984–85

Business Manager
Holly Schauer
P.O. Box 40741
Washington, D.C. 20016

Molecular Developmental Biology

Lawrence Bogorad
Editor

Harvard University
Biological Laboratories
Cambridge, Massachusetts

Alan R. Liss, Inc. • New York

Address all Inquiries to the Publisher
Alan R. Liss, Inc., 41 East 11th Street, New York, NY 10003

Copyright © 1986 Alan R. Liss, Inc.

Printed in the United States of America

Library of Congress Cataloging-in-Publication Data

Main entry under title:

Molecular developmental biology.

"43rd Symposium of the Society for Developmental Biology, Columbia University, June 18–20, 1984, New York."
    Includes index.
    1. Developmental biology—Congresses. 2. Molecular biology—Congresses. I. Bogorad, Lawrence, 1921–   .
QH491.M644   1986   574.3   85-23808
ISBN 0-8451-1504-9

Contents

vi   Contents

Contributors

Gerard F. Barry, Monsanto Company, 700 Chesterfield Village Parkway, St. Louis, MO 63198 **[15]**

Elizabeth H. Blackburn, Department of Molecular Biology, University of California, Berkeley, CA 94720 **[69]**

Lawrence Bogorad, Harvard University, Biological Laboratories, 16 Divinity Avenue, Cambridge, MA 02138 **[ix]**

Ralph L. Brinster, Laboratory of Reproductive Physiology, School of Veterinary Medicine, University of Pennsylvania, Philadelphia, PA 19104 **[149]**

Pierre Chambon, Laboratoire de Génétique Moléculaire des Eucaryotes du CNRS, Unité 184 de Biologie Moléculaire et de Génie Génétique de l'INSERM, Faculté de Médecine, 11 rue Humann, 67085 Strasbourg Cedex, France **[117]**

Howard Y. Chen, Laboratory of Reproductive Physiology, School of Veterinary Medicine, University of Pennsylvania, Philadelphia, PA 19104 **[149]**

Christos Delidakis, Department of Cellular and Developmental Biology, Harvard University, Cambridge, MA 02138 **[85]**

Karl M. Ebert, Laboratory of Reproductive Physiology, School of Veterinary Medicine, University of Pennsylvania, Philadelphia, PA 19104 **[149]**

Jeanne Erickson, Departments of Molecular Biology and Plant Biology, University of Geneva, 30 quai Ernest-Ansermet, CH-1211 Geneva 4, Switzerland **[27]**

Robert T. Fraley, Monsanto Company, 700 Chesterfield Village Parkway, St. Louis, MO 63198 **[15]**

Michel Goldschmidt-Clermont, Departments of Molecular Biology and Plant Biology, University of Geneva, 30, quai Ernest-Ansermet, CH-1211 Geneva 4, Switzerland **[27]**

Leonard Guarente, Department of Biology, Massachusetts Institute of Technology, Cambridge, MA 02139 **[57]**

Muriel Herz, Departments of Molecular Biology and Plant Biology, University of Geneva, 30, quai Ernest-Ansermet, CH-1211 Geneva 4, Switzerland **[27]**

J. Hirsh, Department of Biological Chemistry, Harvard Medical School, 25 Shattuck St., Boston, MA 02115 **[103]**

Robert B. Horsch, Monsanto Company, 700 Chesterfield Village Parkway, St. Louis, MO 63198 **[15]**

The number in brackets is the opening page number of the contributor's article.

Fotis C. Kafatos, Department of Cellular and Developmental Biology, Harvard University, Cambridge, MA 02138, and Institute of Molecular Biology and Biotechnology and Department of Biology, University of Crete, Heraclio, Crete, Greece **[85]**

Katia Komitopoulou, Department of Biochemistry, Cell and Molecular Biology and Genetics, University of Athens, Panepistimiopolis, Kouponia, Athens 15701, Greece **[85]**

Pal Maliga, Advanced Genetic Sciences, Inc., 6701 San Pablo Avenue, Oakland, CA 94608 **[45]**

William Orr, Department of Cellular and Developmental Biology, Harvard University, Cambridge, MA 02138 **[85]**

Richard D. Palmiter, Howard Hughes Medical Institute, Department of Biochemistry, University of Washington, Seattle, WA 98195 **[149]**

Jean-David Rochaix, Departments of Molecular Biology and Plant Biology, University of Geneva, 30, quai Ernest-Ansermet, CH-1211 Geneva 4, Switzerland **[27]**

Stephen G. Rogers, Monsanto Company, 700 Chesterfield Village Parkway, St. Louis, MO 63198 **[15]**

Paolo Sassone-Corsi, Laboratoire de Génétique Moléculaire des Eucaryotes du CNRS, Unité 184 de Biologie Moléculaire et de Génie Génétique de l' INSERM, Faculté de Médecine, 11 rue Humann, 67085 Strasbourg Cedex, France **[117]**

Jeff Schell, Max-Planck-Institut für Züchtungsforschung, D-5000 Köln, Federal Republic of Germany, and Laboratorium voor Genetica, Rijksuniversiteit Gent, B-9000 Gent, Belgium **[3]**

Robert J. Spreitzer, Departments of Molecular Biology and Plant Biology, University of Geneva, 30, quai Ernest-Ansermet, CH-1211 Geneva 4, Switzerland **[27]**

George Thireos, Institute of Molecular Biology and Biotechnology and Department of Biology, University of Crete, Heraclio, Crete, Greece **[85]**

Myrna E. Trumbauer, Laboratory of Reproductive Physiology, School of Veterinary Medicine, University of Pennsylvania, Philadelphia, PA 19104 **[149]**

Jean-Marie Vallet, Departments of Molecular Biology and Plant Biology, University of Geneva, 30, quai Ernest-Ansermet, CH-1211 Geneva 4, Switzerland **[27]**

Marc Van Montagu, Laboratorium voor Genetica, Rijksuniversiteit Gent, B-9000 Gent, Belgium, and Laboratorium voor Genetische Virologie, Vrije Universiteit Brussel, B-1640 St.-Genesius-Rode, Belgium **[3]**

Alan Wildeman, Laboratoire de Génétique Moléculaire des Eucaryotes du CNRS, Unité 184 de Biologie Moléculaire et de Génie Génétique de l'INSERM, Faculté de Médecine, 11 rue Humann, 67085 Strasbourg Cedex, France **[117]**

Yuk-Chor Wong, Department of Cellular and Developmental Biology, Harvard University, Cambridge, MA 02138 **[85]**

Keith R. Yamamoto, Department of Biochemistry and Biophysics, University of California, San Francisco, CA 94143 **[131]**

Preface

The 25th Annual Symposium of the Society was convened in June of 1966 at Haverford, Pennsylvania as a celebration and stock-taking. Its prospective tone was set by its title: "Major Problems in Developmental Biology". Jane M. Oppenheimer, the first speaker of the first session sounded the retrospective chord. Her contribution was entitled "The Growth and Development of Developmental Biology". In this article she reviewed the state of embryology in 1938 and 1939 (the first symposium of the present series was held in August 7–11, 1939 in the Truro Central school at North Truro, Massachusetts). That the study of embryology was the overwhelming component of development in the late 1930's was clear, for the remainder of her discussion of the state of the biology underlying development was lumped under the subtitle "The state of some non-embryological areas of biology from 1938 to 1940." But as Jane Oppenheimer pointed out;

> "The papers in the first symposium [sponsored by the journal "Growth"] were all rather closely related to what in the old days might have been called embryology. This was not to be true for long. As early as 1940, in the second symposium [the first sponsored by the Society for the Study of Development and Growth], every paper dealt with an aspect of what we would call molecular biology. Although most of the speakers . . . addressed themselves to the examination of chemical or physical factors in specific relation to growth and development, O.L. Sponsler in his talk on proteins, and Rudolph Schoenheimer in his on the synthesis of protoplasmic constitutents, made no direct reference to development . . . . "

(To emphasize the antiquity of the term "molecular biology" she goes on elsewhere to point out that W.T. Astbury, in discussing viruses, stated in 1939 that "to the molecular biologist [speaking of himself], the most thrilling discovery of the century is that of the nature of the tobacco mosaic virus; . . . . it is but a nucleoprotein.")

ix

All of this is simply to emphasize that almost from its beginning, the Society for the Study of Development and Growth, which later became the Society for Developmental Biology, has been concerned with the broadest range of biological problems at the level of the single organism from the molecular to the integrated whole. The 43rd Annual Symposium centered on the former with an eye toward the enormous steps forward that knowledge of molecular biology is likely to provide for understanding the development of the latter.

The 41st Symposium held in June 1982, was entitled "Gene Structure and Regulation in Development." Most of the papers dealt with physical characterizations of genes and the monitoring of accumulated RNA and protein products of genes during the development in a variety of organisms—essentially descriptive molecular developmental biology. But among the papers, two pointed to the road leading to the subject of the 43rd Symposium: a paper by Beatrice Mintz entitled "Manipulating the genotype of developing mice" and another by Richard D. Palmiter and Ralph L. Brinster entitled "Inheritable expression of fusion genes microinjected into mouse eggs." In the Society's records these papers mark the sliding transition from descriptive toward experimental molecular developmental biology.

The 43rd Annual Symposium, held at Columbia University in New York June 18–20, 1984 was entitled "Molecular Developmental Biology" but in its earliest, formative times, it had the working subtitle "Watching Foreign Genes at Work." This symposium, with more than 400 people in attendance, examined the state of the opening of the developmental biologist's long dream—a dream that an older generation of developmental biologists hardly imagined having or didn't even know they could have. The dream opens with the possibility of understanding developmental processes by deliberately and specifically altering the genetic make up of an organism and of dissecting the activities of a single gene across developmental time. The sessions of the symposia were entitled Foreign Genes in Eukaryotic Cells; The Expression of Foreign Genes in Plants and Plant Cells; Transformations of Lower Eukaryotes; Transformation Studies Using *Drosophila*; and Vertebrate Transformations. As faint echos of 1982 (only echos because of the exciting new data presented from the same laboratories) Ralph Brinster participated and spoke on "Introduction of new genes into mice" and Beatrice Mintz presented more data along the lines of research she described two years earlier in a paper entitled "Mutagenesis by DNA insertion in developing mice". But the startling difference from a few years earlier was the enormous explosion of transformation studies in cells of plants and animals and of whole organisms. Oh yes, this time things were different from 1940 for the Sponslers and Schoenheimers of 1984 addressed developmental problems!

The breadth of developmental biology and of the interests of members of the Society has been reflected over the years in the subject matter of the Symposia. Until the 10th Symposium in 1951, each symposium was entitled simply "Development and Growth". After that, a single subject was selected for attention. The titles of earlier Symposia are listed elsewhere in this volume. Thus the diversity can be recognized by looking at titles over a number of years but a reflection of the diversity and breadth of Developmental Biology has been provided during the past few Symposia in the posters that are presented. A glimpse of that reflection is provided in this volume in the four abstracts for the prize winning posters presented by graduate students.

The Society is grateful to the National Science Foundation for its financial support of this Symposium. We are also grateful to Joyce E. Jackson, conference housing manager at Columbia University, for her enormous contribution to our intellectual and physical welfare and to Dr. Diane M. Robins of Columbia University who was a one person local committee. Personally, I am grateful to Anne H. Schauer, the Executive Officer of the Society, for her support and assistance during the period that the Symposium was being organized as well as during the meeting itself. I am also indebted to Mr. George Adelman for his assistance in various matters relating to the published proceedings of this Symposium.

Finally, I want to convey my gratitude to those who participated in the Symposium and especially to the speakers who have provided manuscripts which appear here to give those who were not in attendance a feeling for the new avenues opening to students of developmental biology and to remind those who did attend of the continuing excitement of the study.

Lawrence Bogorad

Young Investigators Awards—1984

First Place Award

Vassie C. Ware
Department of Biology
Brown University
Providence, RI 02912

rRNA Processing: Structure of the "Gap" Between α and β 28S
rRNA of Sciara coprophila.

Second Place Awards

Peter A. Merrifield
Department of Biology
University of Virginia
Charlottesville, VA 22901

Myosin Light Chain Switching is Nerve-Dependent in Developing
Limb Musculature

Antonis Hatzopoulos
Department of Biochemistry
Northwestern University
Evanston, IL 60201

Regional Expression of Silkmoth Chorion Genes

David C. Schwartz
Department of Human Genetics and Development
Columbia University
New York, NY 10032

Electrophoretic Separation of Yeast Chromosomal DNAs

Abstract of the First Place
Young Investigator Award—1984

rRNA Processing: Structure of the "Gap" Between α and β 28S rRNA of *Sciara coprophila*. V.C. Ware, R. Renkawitz*, and S.A. Gerbi. Division of Biology and Medicine, Brown University, Providence, RI; *German Cancer Research Center, D-6900, Heidelberg-1, West Germany

In many organisms (e.g., *Protozoa, Mollusca, Annelida, Arthropoda*) mature 28S ribosomal RNA (rRNA) exhibits a central break which divides the rRNA into α and β chains. This rRNA is not colinear with the rDNA, as the "gap" sequence separating the two chains has been removed during RNA processing. Unlike intervening sequences which have been localized in some systems within the β chain coding region, there is no splicing reaction to covalently link the α and β halves together; instead, the two halves are held together by hydrogen bonds. It is unclear why the scission is not a universal phenomenon and how this processing step might alter ribosome function. We have been interested in determining the "gap" structure to learn about its evolutionary origin and possible processing mechanism.

We have determined the sequence of the rDNA region coding for the 28S "gap" in the fungus fly, *Sciara coprophila*, and have used S1 nuclease mapping to define the 5' and 3' boundaries of the "gap". Our data show that only nineteen (19) bases found in rDNA at the "gap" region are absent from mature 28S rRNA. We present a model for the secondary structure of *Sciara* 28S rRNA in the "gap" region based on experiments using S1 nuclease as a probe for single stranded areas in rRNA and reverse transcriptase to direct the synthesis of cDNAs from partially digested rRNAs (Qu *et al.*, 1983. NAR *11*, 5903). By comparison of this model with a model proposed for *Xenopus laevis* 28S rRNA (which lacks a central break) in the counterpart region, clues for processing signals for excision of the "gap" transcript may be revealed.

I. Expression of Foreign Genes in Plants

Molecular Developmental Biology, pages 3–13

Regulated Expression of Foreign Genes of Plants

Jeff Schell and Marc Van Montagu

Max-Planck-Institut für Züchtungsforschung, D-5000 Köln, Federal Republic of
Germany (J.S.), Laboratorium voor Genetica, Rijksuniversiteit Gent, B-9000 Gent,
Belgium (J.S., M.V.M.), and Laboratorium voor Genetische Virologie, Vrije
Universiteit Brussel, B-1640 St.-Genesius-Rode, Belgium (M.V.M.)

I. INTRODUCTION

With the development of efficient gene vector systems for plants, we have
acquired an added experimental capacity to study the molecular organization
of plant genes. In this paper, we review and briefly discuss some recent
progress in research aimed at identifying the regulatory mechanisms that
underly differential and regulated gene expression in plants. The general
strategy has been to identify regulatory sequences by fusing DNA subfrag-
ments suspected to be involved in regulation to DNAs coding for marker
proteins. These chimeric gene constructs are subsequently introduced into
the genomes of plant cell cultures and into fully regenerated plants in which
their expression can be monitored.

The general feasibility of this approach was demonstrated not only by
analogous approaches successfully used in prokaryotes, plants, and animal
cells but specifically for plants by the use of chimeric genes, coding for
enzymes conferring a dominant selectible phenotype on transformed plant
cell cultures and whole plants [Herrera-Estrella et al., 1983a,b; Fraley et al.,
1983; Bevan et al., 1983; De Block et al., 1984; Horsch et al., 1984].

II. PLANT GENE VECTORS

A variety of attempts to transform plant cells using techniques applicable
to animal cell cultures were only marginally successful [Sarkar et al., 1974;

Fernandez et al., 1978; Loesch-Fries and Hall, 1980; Fraley, 1983]. In contrast, the large Ti plasmids carried by *Agrobacteria* have been shown to be very effective gene vectors for a number of plants [for a recent review, see Zambryski et al., 1984; Nester et al., 1984].

Very little is known about the mechanism by which *Agrobacterium* transfers DNA to the plant nucleus. The process must include at least two stages. First, the interaction of *Agrobacterium* with plant cells initiates a series of events, activating specific genes in the bacteria [Matthysse, 1984; Stachel et al., 1984], ultimately leading to the transfer of a segment of the Ti plasmid into the plant cell. Subsequently, a specific fragment, called the *T-DNA*, recombines with nuclear plant DNA resulting in stable integration.

Although the T-DNA is the only portion of the Ti plasmid stably maintained in tumors, none of the genes responsible for plant cell recognition and interaction, and for subsequent steps leading to and including T-DNA integration, are located within this region [Leemans et al., 1982; Zambryski et al., 1983].

When an insertion is made in the T region (e.g., by transposon mutagenesis), the additional DNA is cotransferred and integrated in the plant genome [Hernalsteens et al., 1980]. The T-DNA in this tumor is enlarged by the presence of the transposon, suggesting that a specific mechanism recognizes the boundaries of the T region and integrates the DNA that is situated in between [Lemmers et al., 1980; Holsters et al., 1982].

Molecular cloning and sequencing of the T-DNA/plant junctions from transformed plant cell lines has also revealed that the integration event is rather precise [Thomashow et al., 1980; Zambryski et al., 1980, 1982; Holsters et al., 1982, 1983; Simpson et al., 1982; Yadav et al., 1982]. Comparison of this sequence with the sequences surrounding the left border of the T region in the Ti plasmid reveals at both sides a direct-repeat of 25 bp with some mismatches. This repeat flanks the nopaline T region and also the TL and TR regions of the octopine Ti plasmid [Zambryski et al., 1982; Barker et al., 1983; Gielen et al., 1984].

Genetic analysis of the nopaline Ti plasmid demonstrated the importance of the right 25 bp sequence. Deletions of the left border have very little, if any, effect on T-DNA transfer [Joos et al., 1983]. Elimination of the right border, by contrast, drastically reduces the Ti plasmid's capacity to transfer or insert its T-DNA [Holsters et al., 1980; Joos et al., 1983; Wang et al., 1984]. It is possible that the right end is not only necessary but also sufficient for transfer and/or integration [Caplan et al., 1985]. Indeed, an independent replicon carrying a right border sequence only and a closely linked marker gene will efficiently transfer this gene when complemented in trans by a Ti plasmid with a functional Vir region [Van Haute, 1984].

Furthermore, the border sequences seem to be recognized wherever they are situated in the bacterial genome, in that a T-DNA inserted in the bacterial chromosome is also transferred into the plant cell [Hoekema et al., 1984; Depicker et al., in preparation). However, it is not known whether the border sequences are recognized only in the bacteria or also in the plant cell.

Presentday gene vectors have been developed taking into account that *Agrobacteria* will transfer DNA sequences into plants provided 1) the sequences to be transferred are covalently linked to a 25 bp border or integration sequence and 2) the *Agrobacteria* carry a functional set of the Ti plasmid Vir region genes.

In our group, most of the experiments have been performed with the practical vectors pGV3850 (devoid of any tumor-inducing genes) and pGV3851 (carrying shoot-stimulating genes linked to the transferred sequences) [Zambryski et al., 1983, 1984; De Block et al., 1984].

Two different ways have been developed to introduce foreign DNA into such vectors. The first one is based on the fact that oncogenes of a wild-type Ti plasmid have been replaced by a pBR322 copy [Zambryski et al., 1983]. A foreign gene cloned into pBR322 can be cointegrated with these vectors via homologous recombination. The second system is a binary vector system consisting of a broad host range plasmid carrying T-DNA border sequences and a complementing plasmid containing the Ti Vir functions. To be suitable for different cloning experiments, several unique and convenient restriction sites should lie between the borders. Prototypes of this system have already been shown to work [de Framond et al., 1983; Hoekema et al., 1983]. Current research in our laboratory and other laboratories is directed to designing and testing similar improved plant vectors.

III. EXPRESSION OF TRANSFERRED GENES

Attempts to express directly foreign genes in plants, using the antibiotic resistance genes of prokaryotic plasmids or genes from eukaryotic organisms, such as the yeast alcohol dehydrogenase gene [Barton et al., 1983] or the α-actin and ovalbumin genes from chicken, have failed [Koncz et al., 1984], presumably because plants require specific sequences for transcription initiation. Because of this, it became essential and possible to screen plant DNA sequences for their capacity to direct transcription initiation after transfer into plant genomes.

To do this, "chimeric genes" were constructed consisting of a wide host range constitutive plant promoter sequence such as the 5'-flanking sequence from the T-DNA gene for nopaline synthase linked to the coding sequence of a foreign gene and to appropriate signals for transcription termination. When introduced into plants via the Ti plasmid vectors, the chimeric genes with the coding sequences for, e.g., aminoglycoside phosphotransferase (APH [3'] II from Tn5) or chloramphenicol acetyltransferase (CAT from Tn9) were shown to be expressed [Herrera-Estrella et al., 1983a,b; Bevan et al., 1983; Fraley et al., 1983]. Furthermore, plant cells containing these chimeric genes became resistant to the toxic drugs, indicating that these new constructs could be used for the selection of cells that acquire foreign genes. Indeed, these chimeric genes have been used as dominant selectible markers to select for transformed plant cells [Herrera-Estrella et al., 1983a; Horsch et al., 1984], from which phenotypically normal and fertile plants have been regenerated [De Block et al., 1984]. These plants express the resistance gene in all tissues and in general transmit the introduced gene to their offspring with Mendelian segregation ratios.

The efficiency of transformation obtained by this modified *Agrobacterium* system is so far significantly more efficient and reproducible than other methods used to transfer DNA into plant cells.

IV. LIGHT-INDUCIBLE GENES

The nuclear gene coding for the small subunit (SS) of ribulose 1,5-bisphosphate carboxylase (RuBPcase) is thus far the best analyzed instance of a regulated plant gene.

The RuBPcase consists of eight large subunits of 53,000 MW encoded by the chloroplast genome and of eight small subunits of 14,000 MW encoded by a three- to ten- member nuclear gene family [Berry-Lowe et al., 1982; Cashmore, 1983; Coruzzi et al., 1983; Broglie et al., 1983; Dunsmuir et al., 1983]. The expression of the small subunit gene is regulated in a complex manner. One of the regulating factors is light, which determines the steady-state levels of mRNA [Bedbrook et al., 1980; Smith and Ellis, 1981; Tobin, 1981]. To identify the sequences involved in this regulation, a 973 bp 5'-flanking "promoter" fragment of a pea (*Pisum sativum*) small subunit gene [Cashmore, 1983] was used to construct a chimeric gene in which this promoter fragment was linked to the coding sequence of the bacterial chloramphenicol acetyltransferase (*cat*) gene, which in turn was linked to the 3'-flanking sequence of the nopaline synthase gene [Herrera-Estrella et al., 1984].

It was demonstrated that this chimeric gene, when introduced in tobacco, was expressed. The level of mRNA and of chloramphenicol acetyltransferase activity was compared to that produced by a similar chimeric gene (*nos-cat*) in which the same *cat*-coding sequence was fused to the nopaline synthase promoter sequence [Herrera-Estrella et al., 1983a].

Whereas the *nos-cat* enzyme expressed the *cat* enzyme irrespective of the light conditions under which the tissues were grown or of the differentiation of the tissues or organs analyzed, this was not the case for the *ss-cat* gene. In the latter case, the chimeric gene would only be expressed in tissues harboring fully differentiated chloroplasts when these tissues were grown under light conditions. White tissues kept under light conditions or green tissues kept in the dark did not express the *ss-cat* gene (or did so only weakly).

It is important to note that the conditions required to induce the transformed chimeric *ss-cat* gene are in fact similar to those required to induce the endogenous small subunit gene. These results, therefore, demonstrate that the 5′-flanking "promoter" sequences of the *ss* gene of RuBPcase contain most, if not all, of the sequences necessary to modulate the expression of adjacent coding sequences whether in the normal gene or in experimentally constructed chimeric genes. The light-dependent regulation of these genes is controlled at least partially at the transcriptional level [Gallagher and Ellis, 1982; Herrera-Estrella et al., 1984]. It has been shown that the expression of the *ss* gene is also regulated in a tissue-specific manner; significant levels of SS mRNA are detected only in green leaves and pericarps [Coruzzi et al., 1984]. The question, however, is whether this tissue specificity is directly or indirectly controlled by tissue-specific factors.

To study this question, a cDNA obtained from leaf-specific mRNA and coding for a potato RuBPcase small subunit was used to study the tissue-specific expression of this gene [Eckes et al., 1985]. It was confirmed that the *ss* gene does not require to be in a leaf structure in order to be expressed; it is expressed in green suspension cultures.

It was also shown that the *ss* gene can be expressed in other differentiated tissues, such as roots, provided that these roots do contain fully developed chloroplasts. The level of expression of the SS in illuminated, colorless suspension cultures was very low, and induction was shown to be dependent of the presence of well developed chloroplasts. Clearly several factors are involved that all have to be present in order to get induction: phytochrome, factor(s) correlated with the differentiation of chloroplasts, and light. The observed leaf-specific expression appears to be indirect and owing to the fact that the only tissue that normally combines all the regulatory signals and factors needed for expression of this gene is the leaf. Preliminary evidence indicates that several and possibly most leaf-specific genes are regulated in this indirect fashion. A similar approach has been used [Kaulen et al., in preparation) to study the regulatory sequences involved in the expression of the chalcone synthase (cs) gene, which is the key enzyme in flavonoid biosynthesis (it combines coumaryol-CoA with three malonyl-CoA to form a C-15 chalcone). This gene is regulated at the transcriptional level in cell cultures by light and elicitors [Hahlbrock et al., 1980, 1983].

The 5'-flanking promoter region of the chalcone synthase gene isolated from *Antirrhinum majus* (Sommer and Saedler, personal communication) was fused to the coding sequence of the neomycin phosphotransferase II (*npt-II*) gene of Tn5, and it was shown that this chimeric gene is expressed in tobacco. Whereas the tobacco chalcone synthase gene requires UV light for induction, the *Antirrhinum* gene is induced by normal light. The chimeric *cs-npt*-II gene in tobacco was shown to be induced by UV light in tobacco. These results demonstrate that 5'-flanking sequences of chalcone synthase are responsible for the specific light regulation of this gene in different plants. Work is in progress to pinpoint the precise sequences involved in this light-mediated regulation.

V. TARGETING OF PROTEINS FOR IMPORT IN CHLOROPLASTS

At least 70% of all proteins in chloroplasts are coded for by nuclear genes [Ellis, 1981; Ellis and Robinson, 1984]. These proteins are transported into the plastids by the specific processing of a so-called transit peptide, which normally is an amino terminal extension of the transported protein. Two members of the nuclear-coded chloroplast proteins have been well characterized. These are the chlorophyll a/b-binding proteins and the small subunit of RuBPcase [Chua et al., 1980; Schmidt et al., 1981]. The SS is synthesized on free cytoplasmic ribosomes as a precursor of 20,000 MW and transported into the chloroplasts by a posttranslational and energy-dependent mechanism [Grossman et al., 1982]. The mature polypeptide of the small subunit assembles with the large subunit proteins to form functional RuBPcase present as a soluble protein in the stroma of the chloroplast. The transit sequences of these two systems (SS and chlorophyll a/b-binding protein) in *Pisum sativum* are different in size and do not show any significant homology at the level of their amino acid sequence [Cashmore, 1984], suggesting that a given transit peptide is specific for a given protein being transported.

Furthermore, the SS precursor proteins from soybean, pea, duck weed, and wheat show high homology in a region surrounding the cleavage site of the precursor protein, whereas the homology in the rest of the transit sequences and in the mature small subunit sequence is less pronounced [Broglie et al., 1983]. Nevertheless, small subunit precursors of several higher plants can be transported into each other's chloroplasts [Chua and Schmidt, 1978].

One could therefore expect that a foreign protein similar in size and hydro-phobicity might be transported in chloroplasts when fused to the transit peptide sequence and part of the mature SS protein provided that the entire region of high homology around the cleavage site was present in the fusion protein. Alternatively, one could think that possibly only a transit peptide sequence would be needed to transport a protein into plastids.

These expectations could readily be tested by the strategy of chimeric gene construction and transformation. Two different constructions were made: The first took into account the possibility that the entire region of high homology around the cleavage site would be necessary for transport and processing [Schreier et al., 1985]. A fragment containing the 5'-flanking promoter region as well as the sequences coding for the transit peptide and the first 23 amino acids of the small subunit gene from pea [Cashmore, 1983] was fused to the coding sequence of the *npt*-II gene from Tn5. It was expected that the resulting fusion protein would be active as a neomycin phosphotrans-ferase and would therefore be easily detectible.

In the other construction, the *npt*-II coding sequence was fused directly to the transit peptide sequence such that the potential protein cleavage site did not contain any amino acids derived from the mature small subunit protein [Van den Broeck et al., 1985]. Both chimeric genes were introduced in tobacco and shown to be expressed and to convey kanamycin resistance to the transformed tobacco tissues and plants.

The transcription of the first construction [Schreier et al., 1985] was shown to be light-inducible and to be as efficient as that of the endogenous small subunit gene(s) and more efficient than the expression of the previously described *ss-cat* genes using the same pea small subunit promoter [Herrera-Estrella et al., 1984]. We speculate that the higher level of induced steady-state mRNA in these tissues is owing to improved mRNA stability, possibly resulting from the presence of one intron in the transcript of this transit peptide small subunit *npt*-II chimeric gene *tp-ss-npt*-II) and the absence of any intron in the construction of Herrera-Estrella et al. [1984].

The translation product of the chimeric *tp-ss-npt*-II was shown to be transported to the chloroplasts and to be processed. The data demonstrate that the chimeric *tp-ss-npt*-II gene, which on expression yields a fusion protein with a transit peptide and the conserved amino acid sequence flanking the processing site, is indeed translocated to the chloroplasts and is processed to yield a fusion protein located in the stroma, consisting of the NH_2-terminal end of the small subunit protein and an active NPT-II protein. The results obtained with the other chimeric gene (*tp-npt*-II) [Van den Broeck et al., 1985] demonstrate that the NPT-II component of a precursor protein, which contains only the transit peptide sequences immediately fused to the NPT-II protein and thus missing part of the amino acid sequence flanking the processing site, is equally translocated across the chloroplast envelope and apparently properly processed. The latter results indicate that the transit peptide sequence alone is sufficient to both transport and process precursor proteins into chloroplasts. The function, if any, of the conserved amino acid sequence around the processing site is therefore unclear.

These results demonstrate that it is possible to introduce foreign proteins into plant cell organelles either as such or as fusions with proteins that are coded for by nuclear genes and normally transported into the plant cell chloroplasts. This might lead to a better understanding of the role played by various domains of transported proteins interacting with plastid-coded proteins. The approach described here-in probably does not apply for chloroplasts only but might be generalized to several cellular organelles and compartments. Indeed, Hurt et al. [1984] have independently demonstrated that chimeric genes can be fused to direct the transport of foreign proteins into yeast mitochondria and that the cleavible prepiece of an imported mitochondrial protein is sufficient to direct this transport. Whether or not transit peptides can also be used to direct proteins into membranes such as the thylacoid membrane is unclear. Work is in progress to resolve this question with the transit peptide of the chlorophyll a/b-binding protein (Van den Broeck et al., in preparation).

VI. GENERAL CONCLUSIONS

The use of chimeric genes combined with appropriate gene vector systems has been shown to be a feasible and efficient approach to identifying DNA sequences involved in the regulation of gene expression in plants. Thus such sequences have been found to be located in the 5'-flanking sequences of genes and relatively close to the actual initiation site of transcription. Although our results show that these 5'-flanking sequences play an essential role in determining, e.g., light regulation and tissue-specific expression of plant genes, they do not rule out other important factors such as possible position effects and mRNA stability. The latter factors might very well be determining elements that govern the actual level of steady-state mRNA after induction.

ACKNOWLEDGMENTS

This work was supported by grants from the "A.S.L.K.-Kankerfonds," from the Services of the Prime Minister (O.O.A. 12056184), and from the

"Fonds voor Geneeskundig Wetenschappelijk Onderzoek" (F.G.W.O. 3.001.82) and was carried out under Research Contract No. GVI-4-017-B (RS) of the Biomolecular Engineering Programme of the Commission of the European Communities. GG is Research Assistant to the National Fund for Scientific Research (Belgium).

VII. REFERENCES

Barker RF, Idler KB, Thompson DV, Kemp JD (1983): Plant Mol Biol. 2:335–350.
Barton KA, Binns AN, Matzke AJM, Chilton M-D (1983): Cell 32:1033–1043.
Bedbrook JR, Smith SM, Ellis RJ (1980): Nature 287:692–697.
Berry-Lowe SL, Mc Knight TD, Shah DM, Meagher RB (1982): J Mol Appl Genet 1:483–498.
Bevan MW, Flavell RB, Chilton M-D (1983): Nature 304:184–187.
Broglie R, Coruzzi G, Lamppa G, Keith B, Chua N-H (1983): Biotechnology 1:55–61.
Caplan AB, Van Montagu M, Schell J (1985): J Bacteriol (in press).
Cashmore AR (1983): In T. Kosuge T, Meredith CP, Hollaender A (eds): "Genetic Engineering of Plants—An Agricultural Perspective." New York: Plenum Press, pp 29–38.
Cashmore AR (1984): Proc Natl Acad Sci USA 81:2960–2964.
Chua N-H, Grossman AR, Bartlett SG, Schmidt GW (1980). In Bucher T, Sebald W, Weiss H (eds): "Biological Chemistry of Organelle Formation." Berlin: Springer-Verlag, pp 113–117.
Chua N-H, Schmidt GW (1978): Proc Natl Acad Sci USA 75:6110–6114.
Coruzzi G, Broglie R, Cashmore AR, Chua N-H (1983): J Biol Chem 258:1399–1401.
Coruzzi G, Broglie R, Lamppa G, Chua N-H (1984): In Ciferri O, Dure L III (eds): "Structure and Function of Plant Genomes" (NATO ASI Series, Series A: Life Sciences, Vol 63). New York: Plenum Press, pp 47–59.
De Block M, Herrera-Estrella L, Van Montagu M, Schell J, Zambryski P (1984): EMBO J 3:1681–1689.
de Framond AJ, Barton KA, Chilton M-D (1983): Biotechnology 1:262–269.
Dunsmuir P, Smith S, Bedbrook J (1983): Nucleic Acids Res 11:4177–4183.
Eckes P, Schell J, Willmitzer L (1985): Mol Gen Genet (in press).
Ellis RJ (1981): Annu Rev Plant Physiol 32:111–137.
Ellis RJ, Robinson C (1984): In "The Enzymology of the Posttranslational Modification of Proteins," New York: Academic Press, in press.
Fernandez SM, Lurquin PF, Kado CI (1978): FEBS Lett 87:277–282.
Fraley RT (1983): Plant Mol Biol 2:5–14.
Fraley RT, Rogers SG, Horsch RB, Sanders PR, Flick JS, Adams SP, Bittner ML, Brand, LA, Fink CL, Fry JS, Galluppi GR, Goldberg SB, Hoffmann NL, Woo SC (1983): Proc Natl Acad Sci USA 80:4803–4807.
Gallagher TF, Ellis RJ (1982): EMBO J 1:1493–1498.
Gielen J, De Beuckeleer M, Seurinck J, Deboeck F, De Greve H, Lemmers M, Van Montagu M, Schell J (1984): EMBO J 3:835–846.
Grossman AR, Bartlett SG, Schmidt GW, Mullet JE, Chua N-H (1982): J Biol Chem 257:1558–1563.

Hahlbrock K, Boudet AM, Chappell J, Kreuzaler F, Kuhn DN, Ragg H (1983) In Ciferrio, Dure L III (eds): "Structure and Function of Plant Genomes" (NATO ASI Series, Series A: Life Sciences, Vol 63). New York: Plenum Press, pp 15–23.

Hahlbrock K, Schröder J, Vieregge J (1980): In Fiechter A (ed): "Plant Cell Cultures II" (Advances in Biochemical Engineering, Vol 18). Berlin: Springer-Verlag, pp 39–60.

Hernalsteens JP, Van Vliet F, De Beuckeleer M, Depicker A, Engler G, Lemmers M, Holsters M, Van Montagu M, Schell J (1980): Nature 287:654–656.

Herrera-Estrella L, De Block M, Messens E, Hernalsteens J-P, Van Montagu M, Schell J (1983a): EMBO J 2:987–995.

Herrera-Estrella L, Depicker A, Van Montagu M, Schell J (1983b): Nature 303:209–213.

Herrera-Estrella L, Van den Broeck G, Maenhaut R, Van Montagu M, Schell J, Timko M, Cashmore AR (1984): Nature 310:115–120.

Hoekema A, Hirsch PR, Hooykaas PJJ, Schilperoort RA (1983): Nature 303:179–181.

Hoekema A, Roelvink PW, Hooykaas PJJ, Schilperoort RA (1984): EMBO J 3:2485–2490.

Holsters M, Silva B, Van Vliet F, Genetello C, De Block M, Dhaese P, Depicker A, Inzé D, Engler G, Villarroel R, Van Montagu M, Schell J (1980): Plasmid 3:212–230.

Holsters M, Villarroel R, Gielen J, Seurinck J, De Greve H, Van Montagu M, Schell J (1983): Mol Gen Genet 190:35–41.

Holsters M, Villarroel R, Van Montagu M, Schell J (1982): Mol Gen Genet 185:283–289.

Horsch RB, Fraley RT, Rogers SG, Sanders PR, Lloyd A, Hoffman N (1984): Science 223:496–498.

Hurt EC, Pesold-Hurt B, Schatz G (1984): EMBO J 3:3149–3156.

Joos H, Inzé D, Caplan A, Sormann M, Van Montagu M, Schell J (1983): Cell 32:1057–1067.

Koncz C, Kreuzaler F, Kalman ZS, Schell J (1984): EMBO J 3:1029–1037.

Leemans J, Deblaere R, Willmitzer L, De Greve H, Hernalsteens JP, Van Montagu M, Schell J (1982): EMBO J 1:147–152.

Lemmers M, De Beuckeleer M, Holsters M, Zambryski P, Depicker A, Hernalsteens JP, Van Montagu M, Schell J (1980): J Mol Biol 144:353–376.

Loesh-Fries LS, Hall TC (1980): J Gen Virol 47:323–332.

Matthyse A (1984): In "Genes Involved in Microbe-Plant Interactions" (Advances in Plant Gene Research, Vol 1). Verma DPS, Hohn T (eds): Wien: Springer, pp 33–54.

Nester EW, Gordon MP, Amasino RM, Yanofsky MF (1984): Annu Rev Plant Physiol 35:387–413.

Sarkar S, Upadhya MD, Melchers G (1974): Mol Gen. Genet 135:1–9.

Schreier PH, Seftor EA, Schell J, Bohnert HJ (1985): EMBO J (in press).

Schmidt GW, Bartlett SG, Grossman AR, Cashmore AR- Chua N-H (1981): J Cell Biol 91:468–478.

Simpson RB, O'Hara PJ, Kwok W, Montoya AL, Lichtenstein C, Gordon MP, Nester EW (1982): Cell 29:1005–1014.

Smith SM, Ellis RJ (1981): J Mol Appl Genet 1:127–137.

Stachel S, An G, Nester E (1984): J Cell Biochem 8B [Suppl]: 64.

Thomashow MF, Nutter R, Postle K, Chilton M-D, Blattner FR, Powell A, Gordon MP, Nester EW (1980): Proc Natl Acad Sci USA 77:6448–6452.

Tobin EM (1981): Plant Mol Biol 1:35–51.

Van den Broeck G, Timko MP, Kausch AP, Cashmore AR, Van Montagu M, Herrera-Estrella L (1985): Nature (in press).

Van Haute E (1984): Ph.D. Dissertation, Rijksuniversiteit Gent.

Wang K, Herrera-Estrella L, Van Montagu M, Zambryski P (1984): Cell 38:455–462.

Yadav NS, Vanderleyden J, Bennett DR, Barnes WM, Chilton M-D (1982): Proc Natl Acad Sci USA 79:6322–6326.

Zambryski P, Depicker A, Kruger K, Goodman H (1982): J Mol Appl Genet 1:361–370.

Zambryski P, Herrera-Estrella L, De Block M, Van Montagu M, Schell J (1984): In J. Setlow, Hollaender A (eds): "Genetic Engineering, Principles and Methods, Vol 6." New York: Plenum Press, pp 253–278.

Zambryski P, Holsters M, Kruger K, Depicker A, Schell J, Van Montagu M, Goodman HM (1980): Science 209:1385–1391.

Zambryski P, Joos H, Genetello C, Leemans J, Van Montagu M, Schell J (1983): EMBO J 2:2143–2150.

Molecular Developmental Biology, pages 15–26

Gene Transfer in Plants: A Tool for Studying Gene Expression and Plant Development

Robert T. Fraley, Stephen G. Rogers, Robert B. Horsch, and Gerard F. Barry

Monsanto Company, St. Louis, Missouri 63198

I. INTRODUCTION

The recent development of plant transformation systems has provided a powerful tool for study of plant gene regulation and development. In this chapter the current status of plant cell transformation and production of regenerated plants is reviewed, and the potential use of gene transfer methods in studying light regulation of photosynthetic genes and control of plant development by phytohormones is discussed.

II. PLANT TRANSFORMATION TECHNIQUES: BACKGROUND

The transformation of plant cells by virulent strains of *Agrobacterium tumefaciens* has been studied extensively by several [Chilton et al., 1977; Van Larebeke et al., 1974; Kerr et al., 1977; Braun, 1956] laboratories. A small fragment of the tumor-inducing (Ti) plasmid, called *transferred DNA* (T-DNA), is known to be transferred to and stably incorporated in the nuclear

DNA of transformed plants [Willmitzer et al., 1980; Bevan and Chilton, 1982; Gelvin et al., 1981; Chilton et al., 1980; Yadav et al., 1980], and specific gene products have been shown to be responsible for the observed phytohormone-independent growth characteristics [Leemans et al., 1982; Garfinkel et al., 1981; Willmitzer et al., 1982] and novel metabolic capacities [Holsters et al., 1980] exhibited by crown gall tumor cells. The transfer and insertion of T-DNA into plant DNA is thought to involve repeated nucleotide sequences present near the T-DNA "borders" [Zambryski et al., 1982; Yadav et al., 1982] as well as other genes of unknown function located in specific virulence regions outside T-DNA [Hille et al., 1982; Klee et al., 1982]. The exploitation of the *Agrobacterium*/Ti plasmid system for plant cell transformation has been facilitated by 1) the design of intermediate vectors containing selectable drug markers for introducing foreign genes into the Ti plasmid and subsequently into plant cells, 2) the development of efficient in vitro methods for transforming plant cells and tissues with engineered *Agrobacterium* strains, and 3) the construction of modified *Agrobacterium* strains in which the genes responsible for pathogenicity have been deleted.

III. CONSTRUCTION OF SELECTABLE MARKERS AND INTERMEDIATE VECTORS

To bypass the dependence on tumor genes for identifying transformed plant cells and to overcome the barriers to gene expression in plants, chimeric genes that function as dominant selectable markers have been assembled [Fraley et al., 1983; Bevan et al., 1983; Herrera-Estrella et al., 1983]. These contain the neomycin phosphotransferase (NPT) coding sequences from the bacterial transposon Tn5 (type II) or Tn601 (type I) joined to the 5' and 3' regulatory regions of the nopaline synthase gene from the Ti plasmid.

NPT coding sequences were used in the initial chimeric gene constructions because plant cells were determined to be sensitive to various aminoglycoside antibiotics, and the expression of NPT in yeast [Jiminez and Davis, 1980] and mammalian cells [Colbere-Garapin et al., 1981; Southern and Berg, 1982] has been previously shown to confer resistance to the antibiotic G418. The nopaline synthase gene promoter and 3' nontranslated regions were selected because this gene has been well characterized [Bevan and Chilton, 1982; Depicker et al., 1982], and it is known to be expressed constitutively in most plant tissues transformed with the *A. tumefaciens* Ti plasmid [Tempé and Goldmann, 1982].

Recently, chimeric selectable marker genes have also been constructed using bacterial [Herrera-Estrella et al., 1983] and mouse dihydrofolate reductase (Rogers and Fraley, unpublished observations) and bacterial hygromycin phosphotransferase (Byrne, Rogers, and Fraley, unpublished results) coding sequences.

Because the large size of the Ti plasmid precludes direct cloning approaches, genes are introduced into *A. tumefaciens* cells using intermediate or shuttle vectors. Two types of vectors are typically used, those that integrate via recombination with a resident Ti plasmid [Fraley et al., 1983; Herrera-Estrella et al., 1983] or trans vectors that replicate independently of the Ti plasmid [deFramond et al., 1983; Hoekema et al., 1983; Bevan, 1984]. The integrative vector pMON200 used for the transfer of the chimeric genes into *A. tumefaciens* cells is shown in Figure 1 . Its essential features include 1) a segment of pBR322 DNA for replication in *Escherichia coli*, 2) a segment from pTiT37 that contains a T-DNA border sequence and a functional nopaline synthase gene to facilitate the rapid identification of transformants, 3) a segment of Tn7 carrying the spectinomycin/streptomycin-

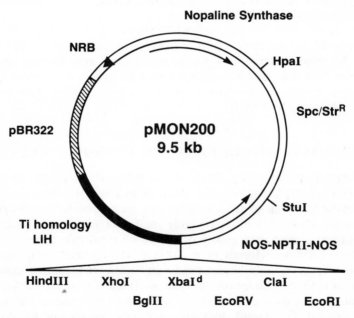

Fig. 1.   The plasmid, pMON200, is a derivative of pMON120 [Fraley et al., 1983] containing a modified chimeric Km^R gene that lacks an upstream ATG codon present in the bacterial leader sequence [Southern and Berg, 1982] and a synthetic multilinker with unique HindIII, XhoI, BglIII, XbaI, ClaI, EcoRI restriction sites. NRB = nopalene T-DNA right border sequence.

resistance determinant for selection in *A. tumefaciens*, 4) a DNA segment obtained from the pTiA6 T-DNA fragment HindIII-18c [see T-DNA map in Leemans et al., 1982], which is included to provide homology for recombination with a resident octopine-type Ti plasmid in *A. tumefaciens*, 5) a synthetic multilinker containing several unique restriction sites for insertion of the chimeric genes, and 6) a chimeric NOS/NPTII/NOS kanamycin-resistance gene. The pMON200 plasmid and derivatives such as pMON128 are introduced into *A. tumefaciens* cells using a conjugation procedure. Homologous recombination between pMON200 or derivatives and a wild type octopine Ti plasmid produces a cointegrate having the structure shown in Figure 2.

The resultant cointegrate plasmid pTiB6S3::pMON128 (Fig. 2C) contains a hybrid T-DNA in which the nopaline-type right border sequence is positioned between the kanamycin-resistance gene and the tumor genes of the resident Ti plasmid. Selection for phytohormone autotrophy results in the "long transfer" event (Fig. 2D) and permits evaluation of the functioning of new marker genes.

Selection for kanamycin resistance and use of the nopaline T-DNA border sequence during infection results in the transfer of a short T-DNA segment (Fig. 2E) that contains the kanamycin-resistance gene and an intact NOS gene but does not contain genes for tumor formation or octopine synthase. The short-transfer transformants can be readily regenerated to give intact plants [Horsch et al., 1984].

IV. CONSTRUCTION OF DISARMED *A. TUMEFACIENS* STRAINS

Although demonstrating the utility of dominant selectable markers for identification of plant cells transformed by avirulent T-DNAs and also permitting the initiation of studies on foreign gene stability and inheritance in plants, the above system is relatively inefficient for routine production of transformed plants; only ∼10% of the transformed calli contain truncated, avirulent T-DNAs [Horsch et al., 1984]. (Preferential utilization of the octopine T_L-DNA right border results in ∼90% of the transformants containing both the phytohormone biosynthetic (tumor) gene and the kanamycin-resistance gene; these calli cannot be regenerated into normal plants.)

A virulent Ti plasmid derivative designed specifically for cointegrate formation with the pMON200-type intermediate vectors and derivatives was constructed as shown in Figure 3A. A double crossover event between the pRK290 derivative and pTiB6S3 (Fig. 3a) results in the deletion of the phytohormone biosynthetic (tumor) genes, the T_L-DNA right right border, and all T_R-DNAs and their replacement with the Tn*903* Km^R marker. Coin-

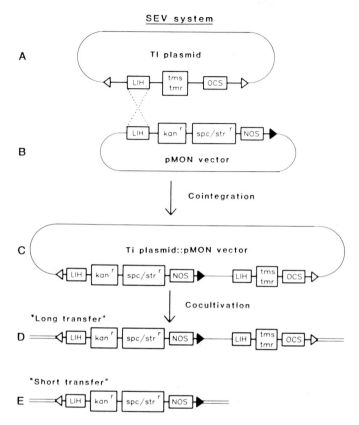

Fig. 2.   Reciprocal recombination of, A, wild type resident Ti plasmid (pTiB6S3) and, B, pMON200 or derivative (pMON128) yields, C, the cointegrate pTiB6S3::pMON128. After cocultivation and selection for kanamycin-resistant plant cells either, D, the entire hybrid T-DNA or, E, a truncated T-DNA without tumor genes is transferred into the plant genome.

tegrate formation between pMON200 and pTiB6S3SE (Fig. 3B) results in the formation of an avirulent, selectable (Km^r) T-DNA. This vector system is referred to as the split end vector (SEV) system because the T-DNA border sequences are present on separate plasmids prior to recombination.

V. IN VITRO PLANT TRANSFORMATION METHODS

A. Cocultivation

In most early studies in plant transformation, transformed plant material was obtained by the infection of whole plants or tissue explants with *A*.

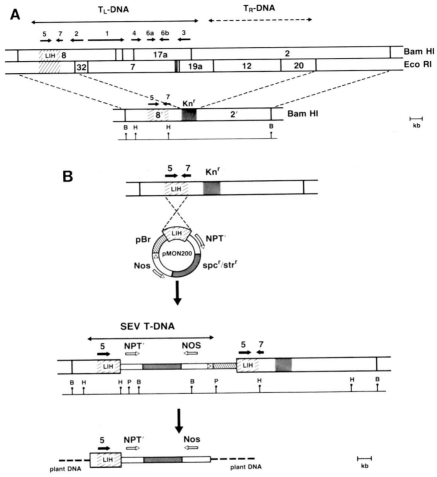

Fig. 3.   Construction of avirulent pTiB6S3SE acceptor plasmid. A, a 4.5 kb BamHI-EcoRI fragment 8': (derived from BamHI fragment 8) and a 6.0 kb EcoRI-BamHI fragment 2': (derived from BamHI fragment 2) were used as homologous DNA segments for double recombination. A 1.2 kb DNA fragment carrying the Km^R determinant from Tn*903* was introduced between the two homologous DNA fragments to provide a marker for subsequent genetic manipulations. The resulting 11.5 kb BamHI fragment was inserted into pRK290 and introduced into *A. tumefaciens* via conjugation. *A. tumefaciens* cells in which the double crossover event had occurred, resulting in the replacement of EcoRI fragments 32, 7, 19a, 12, and 20 with the pMON195 T-DNA, were identified following introduction of the incompatible plasmid pPH1J1. The resulting derivative, pTiB6S3SE, was used as a Ti plasmid acceptor in subsequent experiments with pMON200. B, cointegration between pMON200 and pTiB6S3–SE at their homologous LIH regions results in the formation of a short, avirulent SEV T-DNA containing chimeric Km^R and nopaline synthase genes for monitoring foreign DNA transfer and presence in plant cells. Transfer of the SEV T-DNA into plant cells utilizes the nopaline synthase right border sequence contained in the pMON vector.

tumefaciens cells. Although these procedures are adequate for some purposes, they suffer from the disadvantage of producing relatively small numbers of independent transformants, and they require additional steps to obtain clonal, axenic plant material necessary for many types of analysis. These disadvantages have been overcome by the development of an in vitro transformation procedure in which regenerating protoplasts are cultured directly with *A. tumefaciens* cells [Fraley et al., 1983, 1984; Marton et al., 1979; Wullems et al., 1981]. The bacteria attach to the plant cell walls and presumably introduce T-DNA into the cells in a manner analogous to in planta infection events. Transformed plant cells are readily identified using either kanamycin resistance or phytohormone autotrophy. Transformation frequencies can be as high as 50% using this method.

At present, the in vitro cocultivation method has been successfully applied to *Petunia hybrida* [Fraley et al., 1983], *N. tabacum* [Herrera-Estrella, 1983; Marton et al., 1979; Wullems et al., 1981], *N. plumbaginifolia* (Horsch, et al., 1984) and *H. muticus* [Hanold, 1983] cells. Using this method, transformants can be identified within 3 weeks after protoplast isolation at frequencies near 10^{-1}. Obtaining large numbers of independent transformants will be critical to experiments in which T-DNA vectors are being used for transpositional mutagenesis or "shotgun" cloning approaches or when possible chromosome position effects on gene expression are being examined. Because transformation is likely to occur at the one-or two-cell stage, the resulting transformed calli are predominantly clonal in origin; in most cases, this should eliminate the need for additional single-cell cloning steps.

B. Leaf Disc Transformation

Although the cocultivation procedure has been used quite successfully for initial studies on plant transformation, the requirement for protoplast isolation and regeneration might limit its application with certain crop plants. An alternate method, the leaf disc transformation procedure [Horsch et al., 1985], provides a simple and general method for plant cell transformation, selection, and regeneration that should be applicable to all plant species that can be infected by *Agrobacterium* and regenerated from leaf explants (Fig. 4). Using this procedure, it is possible to obtain transformed shoots within 3–4 weeks after transformation. The ease and reproducibility of this method allow for routine production of transformed plants for studies on plant gene expression and development. Transformed plants produced using either cocultivation [Horsch et al., 1984] or leaf disc transformation methods [Horsch et al., 1985] have been shown to maintain stably and express foreign genes. The progeny from several transformed plants have been shown to inherit foreign DNAs in a Mendelian manner.

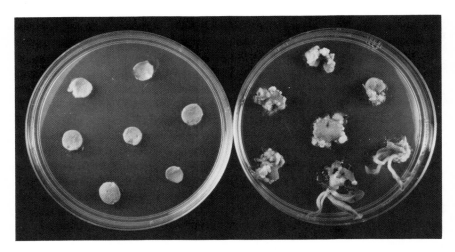

Fig. 4.   Leaf disk transformation and selection of antibiotic-resistant cells. Leaf disks were punched from a surface-sterilized leaf of *Petunia hybrida* (Mitchell), inoculated with *Agrobacterium tumefaciens* strains, cultured on feeder plates, and transferred 2 days later to medium containing 300 μg of kanamycin/ml. The cultures were photographed 21 days after inoculation. Leaf disks transformed with pTiB6S3SE::pMON200 (which contains the chimeric gene for kanamycin resistance) are shown on the right.

VI. USE OF GENE TRANSFER METHODS IN STUDIES OF PLANT GENE EXPRESSION AND DEVELOPMENT

A. Photosynthetic Gene Expression

The availability of reproducible gene transfer methods in plants promises to augment greatly approaches for studying physiological, biochemical, and developmental problems. Analysis of gene expression using deletion and site-directed mutagenesis procedures might allow definition of DNA sequences necessary for tissue-specific and developmentally regulated expression of plant genes. A first example of using gene transfer methods to analyze gene expression in plants is the study of pea ribulose-1,5-bisphosphate carboxylase small subunit gene (pea rbcS gene) expression in petunia [Broglie et al., 1984]. A genomic clone encoding a pea rbcS gene was cloned into a pMON200 derivative vector and following transfer to *A. tumefaciens* cells was introduced into petunia cells via the cocultivation method. Northern S_1 analysis of transformed petunia tissue confirmed that the pea gene was expressed in petunia under the control of its own promoter. In addition, it was demonstrated that expression could be regulated by light, in a fashion identical to that observed for the endogenous rbcS gene in pea. Analysis of regenerated transformed plants indicated that, in addition to normal light regulation, the pea rbcS gene retained its tissue-specific expression pattern in leaves.

Subsequent analysis of deletion mutations in the 5' regulatory region of the pea rbcS gene using transformation assays have revealed that the pea rbcS promoter might contain two functional domains (Morelli and Chua, unpublished observations). A region located 450 bp upstream from the transcription initiation site is necessary for high-level expression, whereas a region between 10 and 30 nucleotides from the transcription start site is necessary for light/dark regulation. Further studies are in progress to define the specified DNA sequence involved in light regulation.

B. Phytohormone Biosynthetic Genes

The role of phytohormones in the regulation of various processes and developmental events has been well documented; however, the molecular basis of phytohormone action is only poorly understood. A powerful approach to studying phytohormone action would be the controlled expression of phytohormone biosynthetic genes in transformed plants. As an initial step in this long-range project, a small region of the Ti plasmid T-DNA region (tmr locus or transcript 4), which was thought to be involved somehow in cytokinin production in metabolism, was cloned and engineered for expression in *E. coli*. By enzyme assay, the *tmr* locus was shown to encode isopentenyl-transferase, the enzyme that catalyzes the first step in cytokinin biosynthesis [Barry et al., 1984]. The product of this reaction is isopentenyl-

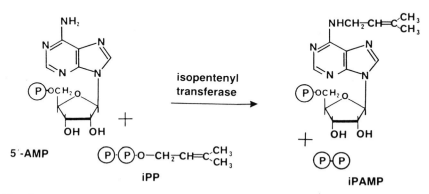

Fig. 5. Reaction catalyzed by isopentenyltransferase: formation of N^6-(Δ^2-isopentenyl) adenosine monophosphate (iPAMP) from Δ^2-isopentenyl (iPP) and adenosine 5' phosphate (5'-AMP).

adenosine monophosphate (iPeAMP; Fig. 5), which by dephosphorylation is converted to the cytokinin isopentenyladenosine (iPeA) and which is the precursor to all other cytokinins.

Recently, biosynthetic genes for IAA production have also been identified [Schröder et al., 1983; Inzé et al., 1984; Thomashow et al., 1984]. The engineering of auxin and cytokinin biosynthetic genes under control of regulated promoters and their reintroduction into plants using gene transfer methods should provide a powerful method for direct analysis of the effects of these phytohormones in plant development.

VII. REFERENCES

Barry, G., Rogers, S., Fraley, R. and Brand, L., Identification of a cloned cytokinin biosynthetic gene, Proc. Natl. Acad. Sci. USA 81, 4776, 1984.

Bevan, M., Binary *Agrobacterium* vectors for plant transformation, Nucleic Acids Res. 12, 8711, 1984.

Bevan, M. and Chilton, M.-D., Multiple transcripts of T-DNA detected in nopaline crown gall tumors, J. Mol. Appl. Genet, 1, 539, 1982.

Bevan, M., Flavell, R.B. and Chilton, M.-D., A chimeric antibiotic resistance gene as a selectable marker for plant cell transformation, Nature 304, 184, 1983.

Braun, A., Plant tumors, Biochim. Biophys. Acta. 516, 167, 1978.

Broglie, R., Coruzzi, G., Fraley, R.T., Rogers, S.G., Horsch, R.B., Niedermeyer, J.G., Fink, C.L., Flick, J.S. and Chua, N.-H., Light-regulated expression of a pea ribulose-1,5-bisphosate carobxylase small subunit gene in transformed plant cells, Science 224, 838, 1984.

Chilton, M.-D., Drummond, M.H., Merlo, D.J. Sciaky, D., Montoya, A.L., Gordon, M.P. and Nester, E.W., Stable incorporation of plasmid DNA into higher plant cells: the molecular basis of crown gall tumorigenesis. Cell 11, 263, 1977.

Chilton, M.-D., Saiki, R.K., Yadav, N., Gordon, M.P. and Quétier, F., T-DNA from *Agrobacterium* Ti plasmid is in the nuclear fraction of crown gall tumor cells. Proc. Natl. Acad. Sci. USA 77, 4060, 1980.

Colbere-Garapin, F., Horodniceanu, F., Kourilsky, P. and Garapin, A.-C., A new dominant hybrid selectable marker for higher eukaryotic cells. Mol. Biol. 150:1–14, (1981).

deFramond, A.J., Barton, K.A. and Chilton, M.-D., Mini-Ti: a new vector strategy for plant genetic engineering, Bio/Technology 1 262, 1983.

Depicker, A., Stachel, S., Dhaese, P, Zambryski, P. and Goodman, H., Nopaline synthase transcript mapping and DNA sequence, J. Mol. Appl. Genet, 1, 561, 1982.

Fraley, R., Horsch, R., Matzke, A., Chilton, M.-D., Chilton, W. and Sanders, P., In vitro transformation of petunia cells by an improved method of cocultivation with *A. tumefaciens* strains, Plant Mol. Biol. 3, 371, 1984.

Fraley, R.T., Rogers, S.G., Horsch, R.B., Sanders, P., Flick, J., Adams, S., Bittner, M., Brand, L., Fink, C., Fry, J., Galluppi, G., Goldberg, S., Hoffmann, N. and Woo, S., Expression of bacterial genes in plant cells, Proc. Natl. Acad. Sci. USA 80, 4803, 1983.

Garfinkel, D.J., Simpson, R.B. Ream, L.W., White, F.F., Gordon, M.P. and Nester, E.W., Genetic analysis of crown gall: find structure map of the T-DNA by site-directed mutagenesis. Cell 27, 143, 1981.

Gelvin, S.B., Gordon, M.P., Nester, E.W., and Aronson, A.A., Transcription of the *Agrobacterium* Ti plasmid in the bacterum and in crown gall tumors. Plasmid 6, 17, 1981.

Harold, D. (1983): *In vitro* transformation of protoplast-derived Hyoscyamus cells by *Agrobacterium tumefaciens*. Plant Sci Lett 30:177–183.

Herrera-Estrella, L., DeBlock, M., Van Montagu, M., Schell, J., Chimeric genes as dominant selectable markers in plant cells, The EMBO Journal, 2, 987, 1983.

Hille, J., Klasen, I., Schilperoort, R. (1982) Construction and application of R prime plasmids, carrying different segments of an octopine Ti plasmid from *Agrobacterium tumefaciens*, for complementation of *vir* genes. Plasmid 7:107–118.

Hoekema, A., Hirsch, P.R., Hooykaas, P.J.J. and Schilperoort, R.A., A binary plant vector strategy based on separation of *vir*- and T-region of the *Agrobacterium tumefaciens* Ti-plasmid, Nature 303, 179, 1983.

Holsters, M., Silva, B., Van Vliet, F., Genetello, C., DeBlock, M., Dhaese, P., Depicker, A., Inźe, D., Engler, G., Villarroel, R., Van Montagu, M. and Schell, J., The functional organization of the nopaline *A. tumefaciens* plasmid pTiC58, Plasmid 3, 212, 1980.

Horsch, R., Fraley, R., Rogers, S., Sanders, P., Lloyd, A. and Hoffmann,W., Inheritance of functional foreign genes in plants. Proc. Natl. Acad. Sci. USA 223, 496, 1984.

Horsch, R.B., Fry, J.E., Hoffmann, N.L., Wallroth, M., Eichholtz, D., Rogers, S.G. and Fraley, R.T., A simple and general method for transferring genes into plants. Science 227, 1229, 1985.

Inźe, E., Folin, A., Van Lijsebettens, M., Simoens, C., Genetello, C., Van Montagu, M. and Schell, J., Genetic analysis of the individual T-DNA genes of *Agrobacterium tumefaciens*; further evidence that two genes are involved in indole-3-acetic acid synthesis. Mol. Gen. Genet. 194, 265, 1984.

Jiminez, A. and Davis, J. (1980) Expression of a transposable element antibiotic resistance element in Saccharomyces. Science 223:496–498.

Kerr, A., Manigault, P., Tempe, J. (1977) Transfer of virulence *in vivo* and *in vitro* in *Agrobacterium*. Nature 265:560–561.

Klee, H.J., Gordon, M.D. and Nester, E.W., Complementation analysis of *Agrobacterium tumefaciens* Ti plasmid mutations affecting oncogenicity. J. Bacteriol. 150, 327, 1982.

Leemans, J., Deblaere, R., Willmitzer, L., De Greve, H., Hernalsteens, J.P., Van Montagu, M. and Schell, J., Genetic identification of functions of T_L-DNA transcripts in octopine crown galls. The EMBO Journal, 1, 147, 1982.

Marton, L., Wullems, G.J., Molendijk, L. and Schilperoort, R.A., In vitro transformation of cultured cells from *Nicotiana tabacum* by Agrobacterium tumefaciens, Nature 277, 129, 1979.

Schröder, G., Waffenschmidt, S., Weiler, E. and Schröder, J., The T-region of Ti plasmids codes for an enzyme synthesizing indole-3-acetic acid, Eur. J. Biochem. 138, 387, 1983.

Southern, P., Berg (1982) Transformation of mammalian cells to antibiotic resistance with a bacterial gene under control of the SV40 early region promoter. J. Mol. Appl. Genet 1:327–341.

Tempé, J., Goldmann, A. (1982): Opine utilization by *Agrobacterium*. In Kahl G, Schell J (eds): "Molecular Biology of Plant Tumors." New York: Academic Press, p 427.

Thomashow, L., Reeves, S., Thomashow, M. (1984): Crown gall oncogenesis: evidence that a T-DNA gene from the *Agrobacterium* Ti plasmid pTiA6 encodes an enzyme that catalyzes synthesis of indoleacetic acid. Proc. Natl. Acad. Sci. USA 81:5071–5076.

Van Larebeke, N., Engler, G., Holsters, M., Van der Elsacker, S., Zaenen, I., Schilperoort, R., Schell, J. (1974) Large plasmids in *Agrobacterium tumefaciens* essential for crown gall inducing ability. Nature 252:169–170.

Willmitzer, L., De Beuckeleer, M., Lemmers, M., Van Montagu, M. and Schell, J., DNA from Ti plasmid present in nucleus and absent from plastids of crown gall plant cells. Nature 287, 359, 1980.

Willmitzer, L., Simons, G. and Schell, J., The T_L-DNA in octopine crown-gall tumors codes for seven well-defined polyadenylated transcripts, The EMBO Journal, 1, 139, 1982.

Wullems, G.J., Molendijk, L., Ooms, G. and Schilperoort, R.A., Retention of tumor markers in F1 progeny plants from *In vitro* induced octopine and nopaline tumor tissues, Cell 24, 719, 1981.

Yadav, N., Postle, K., Saiki, R., Thomashow, M., Chilton, M.-D. (1980) T-DNA of a crown gall teratoma is covalently joined to host plant DNA. Nature 287:458–461.

Yadav, N.S., Vanderleyden, J., Bennet, D., Barnes, W.M., and Chilton, M.-D., Short direct repeats flank the T-DNA on a nopaline Ti plasmid. Proc. Natl. Acad. Sci. USA 79, 6322, 1982.

Zambryski, P., Depicker, A., Kruger, K. and Goodman, H., Tumor induction by *Agrobacterium tumefaciens:* analysis of the boundaries of T-DNA, J. Mol. Appl. Genet. 1, 361, 1982.

Molecular Developmental Biology, pages 27–43

Molecular Genetics of Photosynthesis and Transformation in *Chlamydomonas reinhardii*

Jean-David Rochaix, Jeanne Erickson, Michel Goldschmidt-Clermont, Muriel Herz, Robert J. Spreitzer, and Jean-Marie Vallet

Departments of Molecular Biology and Plant Biology, University of Geneva, CH-1211 Geneva 4, Switzerland

I. INTRODUCTION

One of the distinctive features of eukaryotic cells is their compartmentalized structure. In plant cells, the nucleocytoplasm, the chloroplast, and the mitochondria each contain a distinct genetic system comprising DNA, RNA, ribosomes, and various enzymes and factors involved in DNA replication and transcription and protein synthesis. The existence of these different genetic systems within a single cell implies that their activity must be coordinated. The tight interplay between organellar and nucleocytoplasmic compartments is already apparent in the structure of several organellar proteins consisting of subunits, some of which are encoded and synthesized in one cellular compartment whereas others are encoded and synthesized in the other compartment. Understanding the molecular basis of these multiple interactions remains a challenging problem for the future.

We have deliberately chosen the green unicellular heterothallic alga *Chlamydomonas reinhardii* for studying some aspects of this problem. This alga contains a single large cup-shaped chloroplast that occupies about 40% of the cell volume. Cells of mating type mt^+ and mt^- exist, and they can be propagated vegetatively (Fig. 1). After transfer into a medium lacking a reduced nitrogen source, vegetative cells differentiate into gametes. Fusion of gametes of opposite mating types is followed by nuclear and chloroplast fusion (Fig. 1). The latter property is unique among eukaryotic photosynthetic organisms and has led to extensive studies on chloroplast gene recombination [Sager, 1977; Gillham, 1978]. *C. reinhardii* can be grown either phototrophically in the light with CO_2 on minimal medium lacking a reduced carbon source or on medium containing a reduced carbon source in the light (mixotrophic growth) or in the dark (heterotrophic growth). Mutants deficient in photosynthetic functions can be recognized by their ability to grow under mixotrophic or heterotrophic conditions but not under phototrophic conditions.

The biosynthesis of the photosynthetic apparatus results from a close interplay between the chloroplast and nucleocytoplasmic compartments. An important part of this photosynthetic apparatus is located on the thylakoids, the internal chloroplast membranes, and it consists of an electron transport chain with the two photosystems I and II (Fig. 2). Light energy is captured by the pigment antenna, which consists mostly of chlorophyll, and the energy is channeled to the reaction centers where it induces a charge separation that

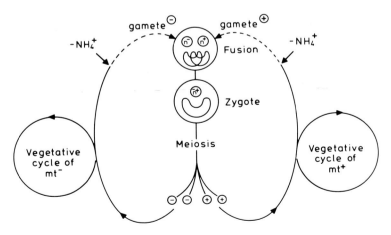

Fig. 1.   Life cycle of *Chlamydomonas reinhardii*. mt, mating type; n, nucleus, see text for explanations.

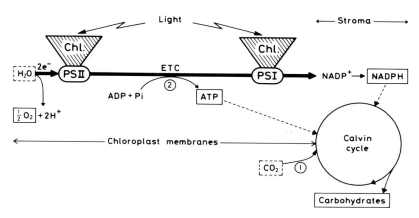

Fig. 2. Scheme of photosynthesis. PSI, PSII, photosystems I, II; Chl, chlorophyll; ETC, electron transport chain. Enzymes 1 and 2 are rubisco and ATP synthase, respectively. See text for explanations.

leads ultimately to the splitting of water into molecular oxygen and protons and electrons. The transport of these electrons along the electron transport chain is coupled with ATP production, and at the end it leads to the reduction of NADP. Both ATP and NADPH are required to drive the reactions of the Calvin cycle, which occur in the soluble phase of the cloroplast in contrast to the primary events of photosynthesis that occur on the thylakoid membranes. CO_2 fixation, one of the major steps of the Calvin cycle, is catalyzed by the enzyme ribulose 1,5 bisphosphate carboxylase/oxygenase (rubisco), the most abundant chloroplast protein.

An important feature is that among the numerous polypeptides involved in photosynthesis some are encoded in the nucleus, translated on cytoplasmic ribosomes, and imported into the chloroplast, and some are coded for by the chloroplast genome and translated on chloroplast ribosomes. The thylakoids of *C. reinhardii* consist of a large number of polypeptides of which at least 40 are synthesized within the chloroplast compartment [Delepelaire, 1984]. The simplest example of a protein with subunits of distinct genetic origin is provided by rubisco whose large and small subunits are encoded by the chloroplast and nuclear genomes, respectively.

Table I summarizes the major properties of the three genetic systems of *C. reinhardii*. Genetic studies indicate the presence of at least 18 linkage groups in the nuclear genome [Harris, 1982]. Although the chloroplast DNA contains only 0.3% of the total cellular information, it constitutes 14% of the cell DNA mass. This implies that the chloroplast DNA, which consists of

TABLE I. Properties of the Genetic Systems of *C. reinhardii*[a]

	Complexity in kb (%)	Mass (%)	Genetics
nuDNA	70 000 (99.7)	(85) Unique	Mendelian
ctDNA	190 (0.3)	(14) 50–80×	UP
mtDNA	15 (0.02)	(1) 50×	BP ≠ Mendelian
	Proteins	rRNAs	Antibiotics
Ribosomes cyt 80S	60	25S,18S,5.8S,5S	chi,spa
ct 70S	50	23S,16S,7S,5S,3S	str,ery,cap

[a]nu, nuclear; ct, chloroplast; mt, mitochondrial; cyt, cytoplasmic; UP, uniparental; BP, biparental; chi, cycloheximide; spa, sparsomycin; str, streptomycin; ery, erythromycin; cap, chloramphenicol. Adapted from Rochaix [1981], with permission.

190 kb circles, is present in 50–80 copies per cell [Behn and Herrmann, 1977; Rochaix, 1978]. Chloroplast genes differ in two important ways from nuclear genes. First, transmission of chloroplast genes is governed by the mating type and occurs uniparentally; i.e., in most cases, only the chloroplast DNA from the mt^+ parent is transmitted to the offspring. Second, in the rare cases of biparental zygotes in which the chloroplast genomes of both parents are transmitted, segregation of chloroplast genes occurs during the post-meiotic mitotic divisions. In contrast to that of higher plants, the mitochondrial genome of *C. reinhardii* has a size of only 15 kb [Grant and Chiang, 1980]. The inheritance of mitochondrial genes appears to be biparental but non-Mendelian [Gillham, 1978]. Table I also indicates that the chloroplast and cytoplasmic ribosomes differ in size, in their proteins, and in the spectrum of antibiotics to which they are sensitive.

The strategy we have used to study chloroplast-nucleocytoplasmic interactions in *C. reinhardii* can be subdivided into three parts. First, the system had to be defined. This part involved the construction of a chloroplast DNA restriction map, the localization of genes of this map, and the structural analysis of several genes involved in photosynthesis from the chloroplast genome and more recently also from the nuclear genome. Second, we have taken advantage of the ability to couple biochemical and molecular analysis with chloroplast genetics in *C. reinhardii* to study the structure-function relationship of chloroplast proteins. Third, we have started to examine the regulation of expression of genes involved in photosynthesis by using specific chloroplast mutants. Attempts to develop an efficient transformation system are in progress, which should allow us to gain further insights into the regulation of this system.

II. DEFINING THE SYSTEM

A restriction map of the chloroplast genome of *C. reinhardii* is shown in Figure 3. A characteristic feature is the inverted repeat, containing the chloroplast rRNA genes, which is also found in most higher plant chloroplast genomes. Numerous genes have been localized on the chloroplast DNA of *C. reinhardii* including the rRNA genes [Rochaix and Malnoe, 1978; Rochaix and Darlix, 1982], tRNA genes [Malnoe and Rochaix, 1978; M. Schneider, unpublished results], and several genes of polypeptides involved in photosynthesis and in chloroplast protein synthesis (see Table II for an exhaustive list). At least two genes, those coding for the 23S rRNA and for a thylakoid polypeptide (psbA in Fig. 3) [Erickson et al., 1984a], contain introns. It is noteworthy that both of these genes are located within the inverted repeat. Sequence analysis of the 5' upstream regions of these chloroplast genes reveals sequence elements that are related to bacterial transcriptional and translational signals [Dron et al., 1982; cf. review of Whitfeld and Bottomley, 1982].

More recently, the nuclear genes of the small subunit of rubisco of *C. reinhardii* have been cloned [Goldschmidt-Clermont, 1984]. There are only two or three copies of these genes in the nuclear genome, clustered at a single locus.

III. TOWARDS CHLOROPLAST MOLECULAR GENETICS

As was mentioned above, *C. reinhardii* is a photosynthetic organism that can be subjected to an extensive genetic analysis both at the nuclear and chloroplast level. We have taken advantage of this important property by examining the molecular basis of two types of chloroplast mutations that map at the rbcL and psbA loci. A list of uniparental mutants that have been characterized at the gene level is given in Table III.

A. rbcL Locus

The rbcL locus codes for the large subunit of rubisco. Several light-sensitive, uniparental rubisco mutants have been characterized. The first isolated mutant, 10-6C, was shown to produce an inactive rubisco enzyme with a lower large-subunit isoelectric point [Spreitzer and Mets, 1980]. Comparison of the sequences of the rbcL genes of this mutant and wild type reveals a single nucleotide change that replaces the gly residue 171 near the first active site of the large subunit with aspartic acid [Dron et al., 1983]. The significance of this base substitution has been confirmed recently when a revertant of this mutant was found to have the original wild-type nucleotide

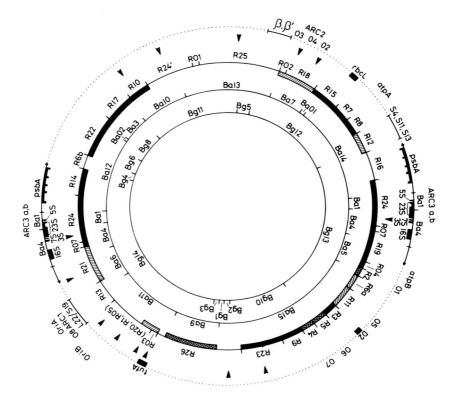

Fig. 3.   Chloroplast DNA map of *Chlamydomonas reinhardii*. The three inner circles from the outside to the inside represent the EcoRI, BamHI, and BglII restriction maps [Rochaix, 1978]. Dark wedges indicate the positions of the 4S RNA genes [Malnoe and Rochaix, 1978]. The two segments of the inverted repeat are drawn on the outside of the map. They contain the rRNA genes and the gene of the 32 kd membrane polypeptide psbA [Erickson et al., 1984a]. The introns in the 23S rRNA gene and in psbA are drawn in thinner lines relative to the coding sequences. D2 is the gene for another photosystem II polypeptide. The genes for the large subunit of ribulose bisphosphate carboxylase, rbcL [Malnoe et al., 1978], for the α and β subunits of the ATP synthase, atpA and atpB, respectively [Woessner et al., 1984; Kovacic and Rochaix, unpublished results], and for the elongation factor EF-Tu, tufA [Watson and Surzycki, 1982] are also indicated. The other gene locations should be considered as tentative; they are based only on heterologous hybridizations with specific probes for the *E. coli* genes of the ribosomal proteins L22 and/or S19, for S4 and/or S11 and/or S13 and for the genes of the β and β' subunits of *E. coli* RNA polymerase [Watson and Surzycki, 1983]. The chloroplast DNA regions whose transcripts are present in large (■), medium (▨), and low (▧) amounts are shown. The eight identified chloroplast ARS sequences are indicated by 01 to 08 [Vallet et al., 1984; Loppes and Denis, 1983]. The four chloroplast DNA sequences promoting autonomous replication in *Chlamydomonas* are marked by *ARC*1, *ARC*2, and *ARC*3a,b [Rochaix et al., 1984a]. Two origins of replication, oriA and oriB, are indicated [Waddell et al., 1984].

TABLE II. Protein Genes Localized on the Chloroplast Genome of *C. reinhardii*

Locus	Polypeptide	Introns	Reference
rbcL	Large subunit of rubisco	0	Malnoe et al. [1979]
psbA	32 kd thylakoid polypeptide associated with PSII, target for herbicides	4	Malnoe et al. [1979]; Erickson et al. [1984]
psbD	D2 polypeptide associated with PSII	0	Rochaix et al. [1984]
atpA	α subunit of ATP synthase	nd[a]	Woessner et al. [1984]; Kovacic and Rochaix (unpublished results)
atpB	β subunit	nd	Woessner et al. [1984]; Kovacic and Rochaix (unpublished results)
atpE	ϵ subunit	nd	Woessner et al. [1984]
atpH	CF_o subunit III	nd	Woessner et al. [1984]
tufA	elongation factor Tu	nd	Watson and Surzycki [1982]

[a]nd, not determined.

TABLE III. Chloroplast Mutations Characterized at the Gene Level

Mutation	Polypeptide affected	Locus	Genetic alteration	Reference
10-6C	Large subunit of rubisco	rbcL	Point mutation	Spreitzer and Mets [1980]; Dron et al. [1983]
18-5B		rbcL	Point mutation	Spreitzer and Ogren [1983]
DCMU4	32 kd membrane polypeptide	psbA	Point mutation	Erickson et al. [1984]
Dr2		psbA	Point mutation	Galloway and Mets [1982]
Ar7		psbA	Point mutation	Galloway and Mets [1984]
11-1A		psbA	Deletion	
8-36C		psbA	Deletion	Spreitzer and Mets [1981]; cf. Figure 4
11-4D		psbA	Deletion	
FuD11-2		psbA	Deletion	
FuD7		psbA	Deletion	Bennoun (unpublished results); cf. Figure 4
FuD13		psbA	Deletion	
FuD-50	β subunit of ATP synthase	atpB	Deletion	Woessner et al. [1982]

[Spreitzer et al., 1982; Spreitzer, Rahire, and Rochaix, unpublished results]. In that the 10-6C mutation has been shown to be linked to other genetic markers in the uniparental linkage group [Mets and Geist, 1983], this mutation provides the first correlation site between the genetic and physical chloroplast DNA maps in *C. reinhardii*. Two other uniparental rubisco mutants, 18-5B and 18-7G, were isolated by screening photosynthetic mutants for their inability to recombine with the original 10-6C mutant [Spreitzer and Ogren, 1983]. Both of these mutants lack rubisco enzyme. Although no large subunit can be detected when cells of 18-7G are labeled for a short period with ^{35}S-sulfate, a slightly truncated product (which is immunoprecipitated with antirubisco antibody) is seen in the 18-5B mutant. Indeed, sequencing of the rbcL gene of this mutant has revealed the presence of a termination codon near the end of the gene. Pulse-chase experiments indicate that the truncated product is unstable. The 18-5B and 18-7G mutants are not only useful for correlating large-subunit structure to function and enzyme assembly and for searching for chloroplast suppressors [Spreitzer et al., 1984], they are also of considerable interest for investigations of the coordination of synthesis of the large and small subunits of rubisco in the chloroplast and nucleocytoplasmic compartments, respectively. Hybridizations of RNA from wild type and from mutant cells with DNA probes specific for the genes of the large and small subunit indicate that these two genes are transcribed at nearly the same level in the mutants and in wild type. Immunoprecipitation of pulse-labeled cells with antirubisco antibody shows that the small subunit is synthesized and processed to its mature size in both mutants, suggesting that it is imported into the chloroplast. It is, however, rapidly degraded in these mutants, indicating that the stoichiometry of the two subunits is achieved at a posttranslational level by some chloroplast-located protease. These results agree with the work of Schmidt and Mishkind [1983], which demonstrated that under conditions of inhibition of chloroplast protein synthesis the small subunit is still synthesized but is rapidly degraded.

B. psbA Locus

The product of the psbA locus has been under intense investigation in several laboratories. This protein, usually referred to as the "32 kd" polypeptide, is associated with photosystem II. It turns over rapidly in the light (i.e., in cells active in photosynthesis) but not in the dark [Reisfeld et al., 1982]. Interestingly, this polypeptide appears to be the target site for several herbicides that block electron transport at the reducing side of photosystem II, presumably by interfering with quinone binding [Arntzen et al., 1982]. Labeled azido atrazine has been shown to bind preferentially to the 32 kd polypeptide [Pfister et al., 1981]. Its amino acid sequence has been highly conserved in higher plants and algae. In contrast to higher plants, in which the psbA gene has been mapped in the single copy region, it is located within the inverted repeat and therefore present in two copies per genome in *C. reinhardii*. As is shown in Figure 4, the gene contains four introns of 1.35, 1.4, 1.1, and 1.8 kb, and it spans a region of 7 kb [Erickson et al., 1984a].

Uniparental mutants resistant to the herbicides diuron, atrazine, and bromacil have been isolated in *C. reinhardii* [Galloway and Mets, 1982, 1984; Tellenbach et al., 1983; Erickson et al., 1984b]. Atrazine resistance has also appeared in several weed species [Arntzen et al., 1983]. The first mutant of *C. reinhardii* examined, DCMU4, was isolated by P. Bennoun as a diuron-resistant mutant. Further studies revealed that this mutant is also highly resistant to atrazine. A comparative sequence analysis of wild-type and mutant psbA revealed a single base pair change in the fifth exon at the ser 264 residue, which is replaced by ala. Both copies of psbA were found to be mutated [Erickson et al., 1984b]. Hirschberg and McIntosh [1983] have previously reported that a similar change occurs in an atrazine-resistant biotype of *Amaranthus hybridus*, in which the same ser is replaced by gly.

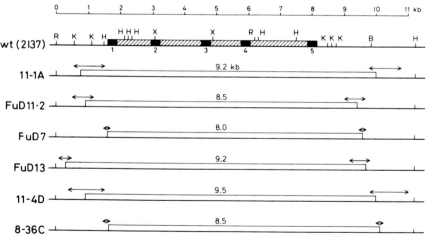

Fig. 4.   Analysis of deletions in the psbA region. The psbA gene is indicated in the upper part of the figure with its five exons (■) and four introns (▨). Transcription proceeds from the left to right. Restriction endonuclease sites are marked by R, EcoRI; K, KpnI; H, HindIII; X, XbaI; B, BamHI. Mutants 11-1A, 11-4D, and 8-36C were isolated by Spreitzer and Mets [1981]; mutants FuD11-2, FuD7 and FuD13 were obtained from P. Bennoun. Deletions are indicated by open bars. The deletion end points were mapped within the regions marked by double arrows.

Sequence analysis of psbA from two other uniparental herbicide-resistant mutants of *C. reinhardii* isolated by Galloway and Mets [1984] has revealed that other parts of the 32 kd polypeptide play a role in herbicide binding. Mutant Dr2, which is 17-fold more resistant to diuron and twofold more resistant to atrazine, has the wild-type val 219 changed to ile; mutant Ar7, which is 15-fold more resistant to atrazine and more sensitive to diuron than wild type, has phe 255 changed to tyr (Erickson, Rahire, and Rochaix, unpublished results). These studies indicate that there is a considerable flexibility in the binding sites for these herbicides. It is noteworthy that electron flow is not affected by the mutations in Dr2 and Ar7, in contrast to DCMU4 and to the atrazine-resistant *Amaranthus*. Mutations of the former type might therefore be of considerable agronomic value.

The tightly packed structural organization of the polypeptides in photosystem II might introduce some complications in the interpretation of these results. The fact that amino acid changes occur in these herbicide-resistant mutants does not necessarily imply that the binding site is located entirely on the psbA product. Allosteric effects on neighboring polypeptides in the photosystem II unit involved in true herbicide binding cannot be ruled out [Gressel, 1984]. It is interesting to note that another PSII-associated polypeptide, D2, displays partial sequence homology with the 32 kd polypeptide, especially in the region where amino acid changes have been found in herbicide-resistant mutants [Rochaix et al., 1984b]. It will be of interest to determine whether or not alterations in this polypeptide can also induce herbicide resistance.

In that the psbA gene spans a relatively large region of the chloroplast genome, one might expect that this gene is particularly sensitive to mutagenesis. Among six mutants examined that are deficient in photosystem II [Spreitzer and Mets, 1981; P. Bennoun, unpublished results] all have deleted the entire psbA region (Fig. 4; Herz, Erickson, and Rochaix, unpublished results). Because the psbA gene is within the inverted repeat, all of these mutants contain double deletions. It appears likely that one deletion was created first and then transmitted to the other copy of the inverted repeat, as seems to be the case for the single-site base substitutions in the herbicide-resistant mutants. The preliminary mapping data are consistent with the deletions being entirely within the inverted repeat. Double deletions at the other end of the inverted repeat have been described previously by Myers et al. [1982]. The mutations affecting psbA are valuable as genetic chloroplast markers. Linkage between the 8-36C mutation (Fig. 4) and other makers of the uniparental linkage group has been demonstrated [Mets and Geist, 1983], and several PSII mutations have been placed in a single chloroplast genetic locus [Spreitzer and Ogren, 1983].

IV. TRANSFORMATION IN *CHLAMYDOMONAS REINHARDII*

The preceding sections have shown how useful defined photosynthetic mutations are in understanding the structure–function relationship of polypeptides and how some of these mutants can be used to examine chloroplast-nucleocytoplasmic cooperation. It is obvious that an efficient transformation system in *C. reinhardii* would be very helpful for developing this analysis further. In that numerous nuclear mutants with deficiencies in their photosynthetic apparatus are available [Harris, 1982], the genes affected could be isolated by complementation through transformation. The possible extension of the transformation to the chloroplast compartment would provide a very powerful tool for understanding the function and regulation of chloroplast genes.

Our first attempts with transformation used the *ARG7* locus as selective marker, because this is one of the few nuclear loci of *C. reinhardii* that has been studied both at the genetic and biochemical levels. The *ARG7* locus codes for arginino succinate lyase (ASL), the last enzyme in the arginine biosynthetic pathway. Several mutants have been isolated at this locus [Gillham, 1965; Loppes et al., 1972], and the enzyme has been partially purified [Matagne and Schlösser, 1977]. A double mutant *cw15, arg7B* was used for most of the experiments. Because the *cw15* mutant is cell-wall deficient [Davies and Plaskitt, 1971] and behaves like a natural protoplast, it is not necessary to use cell wall-degrading enzymes. The yeast *ARG4* locus (which corresponds to the *ARG7* locus of *C. reinhardii*), cloned in plasmic pYe arg4 [Clarke and Carbon, 1978], was used as transforming DNA. After incubation of cells with this DNA in the presence of poly-L-ornithine, colonies able to grow in the absence of arginine were recovered and some of these colonies were shown by Southern hybridization to have the foreign DNA integrated into their nuclear genomes [Rochaix and van Dillewijn, 1982]. A serious limitation was the low transformation frequency, in the range of 10^{-6}–10^{-7} transformants per treated cell, a value only slightly higher than the natural reversion rate. Attempts to clone the *C. reinhardii* *ARG7* locus in yeast by using a cosmid bank of *C. reinhardii* DNA for transformation of yeast *ARG4* strain were not successful. In that this locus appears to be very large based on genetic studies of interallelic complementation [Loppes et al., 1972], it is possible that the ASL gene contains introns that are not processed correctly in yeast.

A limiting step for transformation in this system might occur at the level of stabilization of the foreign DNA once it has entered the cells. One possibility is to use autonomously replicating plasmids as transformation vectors. Because no free plasmids exist in this alga except for chloroplast and mitochondrial DNA, they were constructed in vitro according to the strategy outlined in Figure 5. The 2.7 kb yeast HindIII fragment containing the *ARG*4 locus [Clarke and Carbon, 1978] was inserted into the EcoRI site of pBR322 by blunt-end ligation, thereby producing the plasmid pJD2. MboI and HindIII fragments from total DNA and purified chloroplast DNA were inserted into the BamHI and HindIII sites of pJD2. Pools of these recombinant plasmids were prepared and used to transform yeast or *C. reinhardii* by selecting for arginine prototrophy.

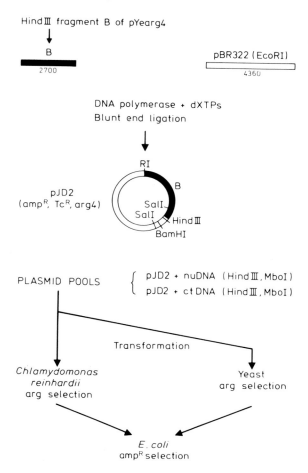

Fig. 5. Strategy for constructing plasmids that replicate autonomously in *C. reinhardii* and yeast. nu DNA, nuclear DNA; ct, chloroplast DNA.

A. Relationship Between Chloroplast ARS Sequences and Authentic Replication Origins of Chloroplast DNA

It was shown previously that a 2.2 kb mitochondrial restriction fragment from *Xenopus laevis* containing the mitochondrial origin of replication promotes autonomous replication in yeast [Zakian, 1981]; i.e., it must contain an ARS element (autonomously replicating sequence). Although this observation did not prove the identity between the ARS element and the mitochondrial origin of replication, it seemed of interest to determine whether or not other organellar origins of replication are related to ARS elements. Pools of recombinant plasmids described above were used for transforming yeast. Most of the transformants contained autonomously replicating plasmids harboring distinct chloroplast DNA segments. Hybridizing these recombinant plasmids to EcoRI- and BamHI-digested chloroplast DNA allowed us to map the chloroplast ARS elements on the chloroplast DNA map (Fig. 3) [Vallet et al., 1984]. To date, our studies and those of Loppes and Denis [1983] have revealed the presence of at least eight chloroplast ARS elements interspersed throughout the chloroplast genome (01 . . . 08 in Fig. 3). Plasmids carrying these elements are unstable in yeast when the cells are grown under nonselective conditions. Sequence analysis of three of these chloroplast ARS fragments reveals a high AT content, many short direct and inverted repeats, and the presence of at least one element in each fragment that is related to the yeast ARS consensus sequence A/T TTT ATPu TTT A/T [Stinchcomb et al., 1981; Broach et al., 1982]. Because of the large number of chloroplast ARS elements, it is very unlikely that they all correspond to authentic chloroplast replication origins. Recently Waddell et al [1984] have mapped two origins of replication of the chloroplast DNA of *C. reinhardii* by observing replication forks in the electron microscope. One of these origins maps on the EcoRI fragment R13, which also acts as an ARS element in yeast. Subcloning of the fragment has shown that the ARS element is distinct from the authentic origin. Similarly we have found that the *Euglena* restriction fragment containing the origin of replication [Koller and Delius, 1982; Schlunegger et al., 1983] does not promote autonomous replication in yeast. These results indicate that at least in *Euglena* and in *C. reinhardii* chloroplast origins of replication are not a subclass of chloroplast ARS elements.

B. Autonomously Replicating Plasmids

Using the pools of recombinant plasmids constructed with total cellular DNA for transformation of *C. reinhardii* cw15 *ARG7*, it was possible to recover several arginine prototrophs. Hybridization of total DNA from these cells with labeled pJD2 DNA indicated the presence of free plasmids [Rochaix et al., 1984a] containing ARC sequences (autonomous replication in *Chlamydomonas*). The amount of plasmid DNA in these cells is considerably lower after 60 generations relative to 25 generations after transformation. Even if the cells are kept under selective pressure, the plasmids are eventually lost, presumably because of reversion at the *ARG7* locus. Several of these plasmids could be recovered by transforming *Escherichia coli* with the DNA isolated from the transformants. Four of these plasmids, pCA1, pCA2, pCA3, and pCA4, were examined in more detail. The locations of the corresponding ARC sequences were determined by hybridizing these plasmids to restriction enzyme digests of nuclear and chloropolast DNA. Surprisingly, all four ARC elements hybridize to chloroplast DNA. Their locations on the chloroplast DNA map are indicated in Figure 3, and their properties are summarized in Table IV. The inserts in these plasmids are relatively small; they range between 102 and 414 bp. Only one of the plasmids, pCA1, has ARS activity. Its insert contains a 27 bp inverted repeat that carries a yeast ARS consensus sequence [Vallet et al., 1984]. It is noteworthy that pCA1 maps on the EcoRI fragment R13, which contains one of the chloroplast origins of replication [Waddel et al., 1984]. Sequence comparison of the four ARC regions has revealed two semiconserved AT-rich sequences of 19 and 12 bp that might play a role in promoting autonomous replication in *C. reinhardii* [Rochaix et al., 1984a].

C. Problems and Prospects

A major problem is that the transformation efficiency is still rather low even with the autonomously replicating plasmids. The limiting step could occur at different levels: delivery of DNA to the cells, stabilization of the foreign DNA, or expression of the selective marker. It is by no means certain that the yeast promoter of the ASL gene is efficiently recognized by the *C. reinhardii* RNA polymerase. Recently the 5′ upstream region of the small subunit of rubisco has been cloned and sequenced (Goldschmidt-Clermont,

TABLE IV. Properties of ARC Plasmids[a]

Plasmid	Site	Size of insert (bp)	Location		ARS activity
pCA1	ARC2	414	R18	Ba7	+
pCA2	ARC1	153	R13	Ba11	−
pCA3	$ARC3_a$	102	R24	Ba4	−
pCA4	$ARC3_b$	257	R24	Ba4	−

[a]R and Ba refer to the chloroplast EcoRI and BamHI fragments shown in Figure 3. ARS activity is defined as the ability to promote autonomous replication in yeast.

unpublished results). Experiments are in progress to put the yeast ARG4 gene under the control of this region.

What are the possibilities of extending the transformation to the chloroplast compartment? Because of the presence of a single large-size chloroplast in *C. reinhardii*, it might be possible to introduce DNA by microinjection into the organelle. The mutants described in a previous section offer interesting possibilities in this respect. Among the rubisco mutants, the most promising mutant appears to be 18-7G; it does not produce detectable levels of rubisco large subunit even during short pulses, and its spontaneous reversion frequency is low. Whether herbicide resistance is a useful transformation marker will depend on whether herbicide resistance is dominant or recessive. The observation that both copies of the psbA gene are mutated in the two herbicide mutants examined is compatible with the latter possibility, although the transmission of one mutation to the other copy might be the result of an efficient mismatch correction mechanism in the inverted repeat or recombination between different chloroplast DNA molecules.

V. CONCLUSIONS

C. reinhardii appears to be an excellent model system for studying the regulation of genes involved in photosynthesis and the integration of organelles within eukaryotic cells, two problems that are fundamental in plant biology. This is owing to the fact that in this organism chloroplast molecular biology and genetics can now be coupled fruitfully at the two loci rbcL and psbA. It is likely that other chloroplast and nuclear loci will be uncovered soon that will allow us to extend this analysis further.

It is possible to introduce foreign DNA into *C. reinhardii*, and it is also feasible to construct plasmids that replicate autonomously in this organism. The development of an efficient transformation system in *C. reinhardii* is an urgent task for the future; it will not only have a great impact on our understanding of regulatory mechanisms in photosynthesis but might also provide a useful test system for genes of higher plants.

ACKNOWLEDGMENTS

We thank O. Jenni for drawings and photography and M. Rahire and J. van Dillewijn for excellent technical assistance. This work was supported by grant 3.258-0.82 from the Swiss National Science Foundation.

VI. REFERENCES

Arntzen CJ, Pfister K, Steinback KE (1982): In Le Baron HM, Gressel J (eds): "Herbicide Resistance in Plants." New York: J. Wiley and Sons, pp 185–214.

Behn W, Herrmann RG (1977): Mol Gen Genet 157:25–30.

Broach JR, Li YY, Feldman J, Jayaram M, Abraham J, Nasmyth KA, Hicks JB (1982): Cold Spring Harbor Symp Quant Biol 47:1165–1173.

Clarke L, Carbon J (1978): J Mol Biol 120:517–532.

Davies DR, Plaskitt A (1971): Genet Res 17:33–43.

Delepelaire P (1984): Photobiochem Photobiophys 6:279–292.

Dron M, Rahire M, Rochaix JD (1982): J Mol Biol 161:775–793.

Dron M, Rahire M, Rochaix JD, Mets L (1983): Plasmid 9:321–324.

Erickson JM, Schneider M, Vallet JM, Dron M, Bennoun P, Rochaix JD (1984a): In Sybesma C (ed): "Advances in Photosynthesis Research, Vol IV." Amsterdam: Martinus Nijhoff, pp 491–500.

Erickson JM, Rahire M, Bennoun P, Delepelaire P, Diner B, Rochaix JD (1984b): Proc Natl Acad Sci USA 81:3617–3621.

Galloway R, Mets L (1982): Plant Physiol 70:1673–1677.

Galloway R, Mets L (1984): Plant Physiol 74:469–474.

Gillham NW (1965): Genetics 52:529–537.

Gillham NW (1978): "Organelle Heredity." New York: Raven Press.

Goldschmidt-Clermont M (1984): In Sybesma C (ed): "Sixth International Congress on Photosynthesis, Vol IV." Amsterdam: Martinus Nijhoff, pp 545–546.

Grant D, Chiang KS (1980): Plasmid 4:82–96.

Gressel J (1984): In Duke SO (ed): "Weed Physiology." Boca Raton, FL: C.R.C. Press, in press.

Harris EH (1982): In O'Brien SJ (ed): "Genetic Maps. Vol II." Frederik, MD: Laboratory of Viral Carcinogenesis, National Cancer Institute, NIH, pp 168–174.

Hirschberg J, McIntosh L (1983): Science 222:1346–1349.

Koller B, Delius H (1982) EMBO J 1:995–998.

Loppes R, Denis C (1983): Curr Genet 7:473–480.

Loppes R, Matagne RF, Strijkert PJ (1972): Heredity 28:239–251.

Malnoe PM, Rochaix JD (1978): Mol Gen Genet 166:269–275.

Malnoe PM, Rochaix JD, Chua NH, Spahr PF (1979): J Mol Biol 133:417–434.

Matagne RF, Schlösser JP (1977): Biochem J 167:71–75.

Mets LJ, Geist LJ (1983): Genetics 105:559–579.

Myers AM, Grant DM, Robert DK, Harris EH, Boynton JE, Gillham NW (1982): Plasmid 7:131–151.

Pfister K, Steinback KE, Gardner G, Arntzen CJ (1981): Proc Natl Acad Sci USA 78:881–885.

Reisfeld A, Mattoo AK, Edelman M (1982): Eur J Biochem 124:125–129.

Rochaix JD (1978): J Mol Biol 126:567–617.

Rochaix JD (1981): Experientia 37:323–332.

Rochaix JD, Darlix JL (1982): J Mol Biol 159:383–395.

Rochaix JD, Dron M, Rahire M, Malnoe PM (1984b): Plant Mol Biol 3:363–370.

Rochaix JD, Malnoe PM (1978): Cell 15:661–670.

Rochaix JD, Rahire M, van Dillewijn J (1984a): Cell 36:925–931.

Rochaix JD, van Dillewijn J (1982): Nature 296:70–72.

Sager R (1977): Adv Genet 19:287–340.

Schlunegger J, Fasnacht M, Stutz E, Koller B, Delius H (1983): Biochim Biophys Acta 478:128–145.

Schmidt G, Mishkind M (1983): Proc Natl Acad Sci USA 80:2632–2636.

Spreitzer RJ, Jordan DB, Ogren WL (1982): FEBS Lett 148:117–121.

Spreitzer RJ, Mets LJ (1980): Nature 285:114–115.

Spreitzer RJ, Mets LJ (1981): Plant Physiol 67:565–569.

Spreitzer RJ, Ogren WL (1983): Proc Natl Acad Sci USA 80:6293–6294.

Spreitzer RJ, Chastain CJ, Ogren WL (1984): Curr. Genet. 9:83–89.

Stinchcomb DT, Mann C, Selker E, Davis RW (1981): ICN-UCLA Symp Mol Cell Biol 22:473–488.

Tellenbach M, Gerber A, Boschetti A (1983): FEBS Lett 158:147–150.

Vallet JM, Rahire M, Rochaix JD (1984) EMBO J 3:415–421.

Waddell J, Wang X-M, Wu M (1984): Nucleic Acids Res 12:3843–3856.

Watson JC, Surzycki SJ (1982): Proc Natl Acad Sci USA 79:2264–2267.

Watson JC, Surzycki SJ (1983): Curr Genet 7:201–210.

Whitfeld PR, Bottomley W (1982): Annu Rev Plant Physiol 34:279–310.

Woessner JP, Masson A, Harris EH, Bennoun P, Gillham NW, Boynton JE (1984). Plant Mol Biol (in press).

Zakian V (1981): Proc Natl Acad Sci USA 78:3128–3132.

Molecular Developmental Biology, pages 45–53
© 1986 Alan R. Liss, Inc.

Cell Fusion to Introduce Genetic Information Coded by Chloroplasts and Mitochondria in Flowering Plants

Pal Maliga

Advanced Genetic Sciences, Inc., Oakland, California 94608

I. INTRODUCTION

This book focuses on genetic modification via transformation, a method not yet applicable to chloroplasts and mitochondria of flowering plants. Owing to their maternal inheritance, these organelles also have not been amenable to classical methods of genetic modification. Traits such as cytoplasmic male sterility, assumed to be controlled by the mitochondrial genome [Leaver et al., 1982], and chloroplast-coded herbicide resistance [Arntzen and Duesing, 1983], cannot be combined by crossing, because chloroplasts and mitochondria are not transmitted through pollen in crop plants [Sears, 1980].

The only way to combine cytoplasmic genetic information in flowering plants has been by somatic cell fusion. In this chapter, the principles of the manipulation of chloroplasts and mitochondria by this method will be illustrated, using mainly *Nicotiana*, a genus in which these problems have been studied most intensively. Fusion of somatic cells, unlike fertilization, results in cells with a mixed population of organelles. In each cell, there are several chloroplasts and mitochondria [Gillham, 1978]. Theoretically, a new cytoplasm can be obtained by combining the intact chloroplasts of one line with intact mitochondria of another. Alternatively, genetic recombination between parental chloroplast or mitochondrial genomes may occur in heteroplasmic cells that contain a mixed population of organelles. Results and problems of isolating lines with novel cytoplasms will be reviewed.

Both cells involved in fusion contribute their nuclei to the heterokaryons. Elimination of one of the nuclei is usually desirable, e.g., when transferring a cytoplasmic element from a weed species into a cultivated line. Practical ways of eliminating one of the nuclei in heterokaryons will be discussed.

Many of the examples presented derive from research carried out at the Institute of Plant Physiology, Biological Research Center, Szeged, Hungary. No attempt will be made to include all results from other laboratories.

II. ORGANELLE MARKERS

The development of various methods of organelle manipulation has been aided by the availability of selectable cytoplasmic markers such as resistance to antibiotics. Markers that are not expressed in tissue culture have also been used in some experiments.

A. Chloroplast Markers

Resistance to streptomycin [Maliga et al., 1973, 1975] and lincomycin [Cseplo and Maliga, 1982, 1984] are defined as the ability to form a green callus on a selective medium that would normally inhibit greening. Sensitive calli are bleached on selective media, but they survive and proliferate at a slower rate. Survival of sensitive calli is probably crucial for the recovery of the mutants themselves, and for the recovery of chloroplast recombinants, by providing sufficient time for the altered chloroplast genomes to become predominant through a sorting out process.

That the chloroplast genome is the location of the antibiotic resistance factors has been inferred from experiments in which the expression of the antibiotic-resistance phenotypes has been correlated with the presence of chloroplasts from the resistant line [Menczel et al., 1981, 1982, 1983; Maliga et al., 1982; Cseplo et al., 1984]. The possibility that mitochondria are involved in determining the resistance phenotypes could not be excluded [Menczel et al., 1983]. For a more detailed discussion of the antibiotic resistance markers in flowering plants, see the review by Maliga [1984].

As nonselective chloroplast markers, DNA restriction site polymorphism [e.g., Belliard et al., 1978; Medgyesy et al., 1985], cytoplasmic pigment mutations [Sidorov et al., 1981; Gleba et al., 1984], and the polypeptide composition of fraction I protein [Chen et al., 1977] have been used.

B. Mitochondrial Markers

No selectable antibiotic-resistance mutations have been assigned yet to the mitochondrial genome. The nonselective markers include cytoplasmic male sterility, a trait expressed only in regenerated plants [Belliard et al., 1979; Galun et al., 1982; Izhar et al., 1983; Menczel et al., 1983], and restriction site polymorphism [Belliard et al., 1979; Nagy et al., 1981].

III. CYBRIDS AND SOMATIC HYBRIDS FROM HETEROKARYOTIC CELLS

Fusion of protoplasts results in mixing of the cytoplasms of the two cells while the nuclei remain separated. In the heterokaryon, the nuclei may fuse by some mechanism to give a somatic hybrid. If the nuclei segregate, the resulting cells will be identical to the parental type with respect to their nuclear genetic material. Because the cytoplasm will be mixed, however, there is chance for hybrid cytoplasm formation, in which case the lines are termed *cybrids*. Cybrids are desirable if introduction of cytoplasmic genetic information, but not of nuclear genetic information (chromosomes), is required. Somatic hybrids, unlike sexual hybrids, can also have a hybrid cytoplasm.

The frequency of nuclear fusion and nuclear segregation in heterokaryons has been studied. Heterokaryon-derived clones were identified by the expression of a cytoplasmic mutation rescued from inactivated protoplasts. In one case, described by Medgyesy et al. [1980], protoplasts of the cytoplasmic streptomycin-resistant mutant of *Nicotiana tabacum* SR1 were treated with iodoacetate to prevent cell division. They were then fused with *Nicotiana sylvestris* (streptomycin-sensitive) protoplasts. Clones expressing streptomycin resistance as the result of cytoplasmic mixing and rescue of the resistant organelles have been identified in a selective medium, and the proportion of somatic hybrids and cybrids has been determined. In about 80% of the clones, the nuclei fused as inferred from regenerating somatic hybrids from the resistant calli. In the rest of the clones, cybrids were obtained as the result of the separation of the parental nuclei. In some clones, both cybrids and somatic hybrids regenerated, suggesting the fusion of more than two protoplasts. It is important to note that the iodoacetate treatment did not induce genetic changes as far as could be determined studying a relatively small number of regenerated plants [Medgyesy et al., 1980].

In a different species combination (*N. tabacum* SR1 + *Nicotiana plumbaginifolia*) the nuclei fused, yielding somatic hybrids in 98% of the clones identified by rescue of the streptomycin-resistance marker from iodoacetate-treated protoplasts [Menczel et al., 1982]. If, however, the streptomycin-resistant SR1 protoplasts were inactivated by Co^{60} irradiation, the regenerated plants in 80% of the clones contained the nucleus of the nonirradiated *N. plumbaginifolia* parent. The elimination of the irradiated nucleus, when it occurred, seems to have been complete. The irradiated nucleus has also been rescued via fusion, and somatic hybrids were regenerated in at least 20% of the clones. The value of irradiation as a tool to facilitate the elimination of the irradiated nucleus has been confirmed by transferring cytoplasmic male sterility [Menczel et al., 1983] and cytoplasmic lincomycin resistance [Cseplo et al., 1984] without transferring nuclear genetic material from the cytoplasmic donor.

Irradiation and iodoacetate treatments can be combined in such a way that only heterokaryon-derived clones are able to form a colony [Sidorov et al., 1981]. Using this technique, efficient organelle transfer can be carried out in the absence of selectable nuclear or cytoplasmic genes.

IV. CHLOROPLASTS IN HETEROPLASMIC CELLS

A. Chloroplast Segregation

Mixed chloroplast populations obtained by somatic cell fusion quickly segregate. Chloroplast segregation in the fusion-derived cells has been random, complete, and relatively quick in the absence of selection pressure [Chen et al., 1977; Belliard et al., 1978; Scowcroft and Larkin, 1981; Schiller et al., 1982]. Still, variegated sectors have frequently been observed on the leaves of regenerated plants when one of the parental lines carried a chloroplast-coded pigment mutation [Gleba et al., 1984]. In rare cases, the heteroplastidic condition has been transmitted to the seed progeny of regenerated plants [Fluhr et al., 1983].

When streptomycin or lincomycin selection has been applied in marker rescue experiments, chloroplasts were present only from the resistant parent in regenerated plants [Menczel et al., 1983; Cseplo et al., 1984]. This might be owing to the competitiveness of the cells with a resistant phenotype that is manifested as faster growth on a selective medium [Maliga et al., 1975]. Maintenance of heteroplastidic condition in the absence of streptomycin selection and rapid elimination of sensitive chloroplasts on a selective medium have been shown [Fluhr et al., 1983].

B. Chloroplast Recombination

Chloroplast DNA in the cybrids and somatic hybrids, when tested for restriction site polymorphism, was identical with one or the other parental type. This means that the chloroplasts segregated as intact, independent units, and no genetic recombination occurred between the two parental chloroplast genomes. This behavior, as will be discussed in the next section, is very different from the behavior of mitochondria.

Absence of chloroplast recombinants in earlier experiments suggested that, if recombination occurs at all, it would be rare. A *N. tabacum* strain with multiple chloroplast markers has been constructed to enable the detection of rare recombinants. This line, SR1-A15, is a derivative of the SR1 mutant and carries an additional, maternally inherited pigment mutation that prevents the expression of streptomycin resistance.

Selection of putative chloroplast recombinants was based on the expression of streptomycin resistance as the result of the correction of the pigment mutation via recombination with the chloroplast genome of a streptomycin-sensitive *N. plumbaginifolia*. Mapping of the chloroplast genome in one of the putative recombinants has revealed alternating sequences from both parental chloroplast genomes [Medgyesy et al., 1985].

V. MITOCHONDRIA IN HETEROPLASMIC CELLS

A. Mitochondrial Segregation

Segregation of intact mitochondria has not been studied, because no parental mitochondrial genomes have been recovered from fusions in which the parental genomes are distinguishable by restriction site polymorphism (see IV.B.). Segregation of a mitochondrial trait, cytoplasmic male sterility, however, was studied in regenerated plants and their seed progeny. In *Nicotiana* cybrids, segregation is complete by the time the regenerated plants flower [Belliard et al., 1978; Aviv and Galun, 1980] except in rare instances [Menczel et al., 1983]. In *Petunia*, the heteroplasmic condition seems to be maintained longer, in some cases through several meiotic cycles [Izhar et al., 1983].

B. Mitochondrial Recombination

Changes in mitochondrial DNA restriction patterns have been discovered early in *Nicotiana* plants regenerated after fusing heteroplasmic lines, that is, after fusing lines in which the mitochondrial genomes could be distinguished by restriction site polymorphism [Belliard et al., 1979; Nagy et al., 1981; Galun et al., 1982]. Similar observations have been made in *Petunia* plants

[Boeshore et al., 1983]. In that not all parent-specific fragments were present in the patterns and a few non-parental fragments were found, the novel patterns could not be owing to the persistence of a mixed parental mitochondrial population. Nonparental mitochondrial DNA patterns have been interpreted as being the result of genetic recombination between the parental mitochondrial genomes [Belliard et al., 1979; Galun et al., 1982]. Segregation of the heterogenous parental mitochondrial DNA molecules was suggested as an additional mechanism that could contribute to the variability [Nagy et al., 1981].

Mitochondrial DNA has also been studied in lines derived from homoplasmic fusions in which the mitochondrial restriction patterns of the parents were identical. The mitochondrial DNA in these lines was parental, whereas it was nonparental in all lines derived from heteroplasmic fusions [Nagy et al., 1983]. This might be explained by the lack of recombination in the homoplasmic combinations or by recombination occurring only through homologous sequences, as has been postulated for corn [Lonsdale et al., 1983] and *Brassica* [Palmer and Shields, 1984].

Changes in the mitochondrial genome of plants regenerated from tissue culture have been observed in corn [Gengenbach et al., 1981; Umbeck and Gengenbach, 1983] and potato [Kemble and Shepard, 1984]. Ten antibiotic-resistant mutants isolated in *Nicotiana* protoplast cultures have therefore been tested to determine to what extent the variability observed in the fusion-derived clones could be caused by the cells being exposed to tissue culture conditions. Patterns in each of the mutants were identical to the wild type [Nagy et al., 1983]. Tissue culture-induced genetic variability could therefore be excluded as the cause of the observed changes in the *Nicotiana* cybrids and somatic hybrids.

VI. CONCLUSIONS

Cells of flowering plants contain several to many chloroplasts and mitochondria. Nevertheless, using appropriate genetic techniques, rare events such as new chloroplast mutations and recombinant chloroplast genomes can be identified. As a result of organelle segregation, these rare genotypes can easily be obtained in pure populations. In crops for which there is technology for plant regeneration from cultured cells the breeding of cytoplasmic organelles has become a reality.

The experiments described have been carried out in sexually compatible species combinations. The next step should be to attempt marker exchange between sexually incompatible species to test the limits within which economically important traits coded for by a chloroplast gene, such as herbicide resistance, can be spread by recombination. Identification of genes controlling functions like cytoplasmic male sterility is also possible now by mapping mitochondrial recombinants and correlating the maps with the flower phenotype.

VII. REFERENCES

Arntzen CJ, Duesing JH (1983): Chloroplast-encoded herbicide resistance. In Downey K, Voellmy RW, Ahmad F, Schultz J (eds): "Molecular Genetics of Plants and Animals." New York: Academic Press, pp 273–292.

Aviv D, Galun E (1980): Restoration of fertility in cytoplasmic male sterile (CMS) *Nicotiana sylvestris* by fusion with X-irradiated *N. tabacum* protoplasts. Theor Appl Genet 58:121–127.

Belliard G, Pelletier G, Vedel F, Quetier F (1978): Morphological characteristics and chloroplast DNA distribution in different cytoplasmic parasexual hybrids of *Nicotiana tabacum*. Mol Gen Genet 165:231–237.

Belliard G, Vedel F, Pelletier G (1979): Mitochondrial recombination in cytoplasmic hybrids of *Nicotiana tabacum* by protoplast fusion. Nature 281:401–403.

Boeshore ML, Hanson MR, Lifshitz I, Izhar S (1983): Novel composition of mitochondrial genomes in *Petunia* somatic hybrids derived from cytoplasmic male sterile and fertile plants. Mol Gen Genet 190:459–467.

Chen K, Wildman SG, Smith HH (1977): Chloroplast DNA distribution in parasexual hybrids as shown by polypeptide composition of fraction I protein. Proc Natl Acad Sci USA 74:5109–5112.

Cseplo A, Maliga P (1982): Lincomycin resistance, a new type of maternally inherited mutation in *Nicotiana plumbaginifolia*. Curr Genet 6:105–109.

Cseplo A, Maliga P (1984): Large scale isolation of maternally inherited lincomycin resistance mutations in diploid *Nicotiana plumbaginifolia* protoplast cultures. Mol Gen Genet 196:407–412.

Cseplo A, Nagy F, Maliga P (1984): Interspecific protoplast fusion to rescue a cytoplasmic lincomycin resistance mutation into fertile *Nicotiana plumbaginifolia* plants. Mol Gen Genet 198:7–11.

Fluhr R, Aviv D, Edelman M, Galun E (1983): Cybrids containing mixed and sorted-out chloroplasts following interspecific somatic fusion in *Nicotiana*. Theor Appl Genet 65:289–294.

Galun E, Arzee-Gonen P, Fluhr R, Edelman M, Aviv D (1982): Cytoplasmic hybridization in *Nicotiana*: Mitochondrial DNA analysis in progenies resulting from fusion between protoplasts having different organelle constitutions. Mol Gen Genet 186:50–56.

Gengenbach RB, Connelly JA, Pring DR, Conde MF (1981): Mitochondrial DNA variation in maize plants regenerated during tissue culture selection. Theor Appl Genet 59:161–167.

Gillham NW (1978): "Organelle Heredity." New York: Raven Press.

Gleba Y, Kolesnik NN, Meshkene IV, Cherep NN, Parokonny AS (1984): Transmission genetics of the somatic hybridization process in *Nicotiana* 1. Hybrids and cybrids among the regenerates from cloned protoplast fusion products. Theor Appl Genet 69:121–128.

Izhar S, Schlicter M, Swartzberg D (1983): Sorting out of cytoplasmic elements in somatic hybrids of *Petunia* and the prevalence of the heteroplasmon through several meiotic cycles. Mol Gen Genet 190:468–476.

Kemble RJ, Shepard JF (1984): Cytoplasmic DNA variation in a potato protoclonal population. Theor Appl Genet 69:211–216.

Leaver CJ, Forde BG, Dixon LK (1982): Mitochondrial genes and cytoplasmically inherited variation in higher plants. In Slonimski P, Borst P, Atardi G (eds): "Mitochondrial Genes." Cold Spring Harbor, New York: Cold Spring Harbor Laboratory, pp 457–470.

Lonsdale DM, Hodge TP, Fauron CM-R, Flavell RB (1983): A. predicted structure for the maize mitochondrial genome from the fertile cytoplasm of maize. In Goldberg RB (ed): "Plant Molecular Biology." New York: Alan R. Liss, Inc., pp 445–456.

Maliga P (1984): Isolation and characterization of mutants in plant cell culture. Annu Rev Plant Physiol 35:510–542.

Maliga P, Breznovits A, Marton L (1973): Streptomycin resistant plants from haploid callus culture of tobacco. Nature [New Biol] 244:28–30.

Maliga P, Breznovits A, Marton L (1975): Non-Mendelian streptomycin resistant mutant with altered chloroplasts and mitochondria. Nature 255:401–408.

Maliga P, Lorz H, Lazar G, Nagy F (1982): Cytoplast-protoplast fusion for interspecific chloroplast transfer in *Nicotiana*. Mol Gen Genet 185:211–215.

Medgyesy P, Fejes E, Maliga P (1985): Interspecific chloroplast recombination in a *Nicotiana* somatic hybrid. Proc Natl Acad Sci USA (in press).

Medgyesy P, Menczel L, Maliga P (1980): The use of cytoplasmic streptomycin resistance: chloroplast transfer from *Nicotiana tabacum* into *Nicotiana sylvestris* and isolation of their somatic hybrids. Mol Gen Genet 179:693–698.

Menczel L, Galiba G, Nagy F, Maliga P (1982): Effect of radiation dosage on efficiency of chloroplast transfer by protoplast fusion in *Nicotiana*. Genetics 100:187–195.

Menczel L, Nagy F, Kiss ZS, Maliga P (1981): Streptomycin resistant and sensitive somatic hybrids of *Nicotiana tabacum* + *Nicotiana knightiana*: Correlation of resistance of *N. tabacum* plastids. Theor Appl Genet 59:191–195.

Menczel L, Nagy F, Lazar G, Maliga P (1983): Transfer of cytoplasmic male sterility by selection for streptomycin resistance after protoplast fusion in *Nicotiana*. Mol Gen Genet 189:365–369.

Nagy F, Lazar G, Menczel L, Maliga P (1983): A heteroplasmic state induced by protoplast fusion is a necessary condition for detecting rearrangements in *Nicotiana* mitochondrial DNA. Theor Appl Genet 66:203–207.

Nagy F, Torok I, Maliga P (1981): Extensive rearrangements in the mitochondrial DNA in somatic hybrids of *Nicotiana tabacum* and *Nicotiana knightiana*. Mol Gen Genet 183:437–439.

Palmer JD, Shields CR (1984): Tripartite structure for the *Brassica campestris* mitochondrial genome. Nature 307:437–440.

Schiller B, Herrmann RG, Melchers G (1982): Restriction endonuclease analysis of plastid DNA from tomato, potato and some of their somatic hybrids. Mol Gen Genet 186:453–459.

Scowcroft WR, Larkin PJ (1981): Chloroplast DNA assorts randomly in intraspecific somatic hybrids of *Nicotiana debney*. Theor Appl Genet 60:179–184.

Sears BB (1980): Elimination of plastides during spermatogenesis and fertilization in the plant kingdom. Plasmid 4:233–255.

Sidorov VA, Menczel L, Maliga P (1981): Chloroplast transfer in *Nicotiana* based on metabolic complementation between irradiated and iodoacetate treated protoplasts. Planta 152:341–345.

Umbeck PF, Gengenbach BG (1983): Reversion of male-sterile T-cytoplasm in maize to fertility in tissue culture. Crop Sci 23:584–588.

II. Gene Expression Control in Lower Eukaryotes

Molecular Developmental Biology, pages 57–68
© 1986 Alan R. Liss, Inc.

Regulation of Cytochrome-Encoding Genes in *Saccharomyces cerevisiae*

Leonard Guarente

Department of Biology, Massachusetts Institute of Technology, Cambridge, Massachusetts 02139

I. INTRODUCTION

Developmental processes are likely to be controlled by the coordinate expression of cell type-specific gene families of an organism. An understanding of the molecular events that underlie coordinate gene control in eukaryotic cells is thus of central interest to developmental and molecular biologists alike. In invertebrate systems, such as *Drosophila*, insight into global systems that coordinate development can be provided by the homeotic mutants that perturb the normal pattern of segmental development of the fly. The simplest view would posit that the homeotic genes encode products that regulate other genes involved in the development of particular body segments.

The lower eukaryotes provide yet simpler systems in which studies of global control systems are accessible to genetic and molecular analysis. Such global systems include, for example, gene families that determine the biogenesis of specific subcellular structures or organelles. Subsets of the intact organelle that would be simple enough to permit experimental analysis are biochemical pathways housed in a particular cellular compartment. An example of such a pathway that we have focused on is the set of cytochromes of the electron transport chain in the inner mitochondrial membrane (Fig. 1). Genes encoding certain of these cytochromes (c1, c, and, four of seven

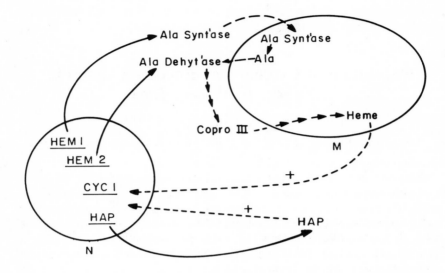

Fig. 1. The biosynthetic pathway for the production of heme for cytochromes. *HEM1*, the gene coding for aminolevulinate synthetase; *HEM2*, the gene encoding aminolevulinate dehydratase; *CYC1*, the gene encoding apocytochrome cl; *HAP*, loci and their products that regulate cytochrome expression via UAS (upstream activating sequences); Copro III, copro-porphyrin III; Ala, aminolevulinate; N, nucleus; M, mitochondrion.

subunits of cytochrome c oxidase) are nuclear, whereas genes encoding the balance of the cytochromes (b and the remaining three subunits of oxidase) lie in the genome of the mitochondria [Katan et al., 1976; Lin and Beattie, 1976; Mason and Shatz, 1973]. In the yeast *Saccharomyces cerevisiae*, there is the added complexity that two distinct genes encode two forms of cytochrome c. *CYC1* encodes the iso-1-cytochrome c, which makes up about 95% of the cellular cytochrome c under most growth conditions, and *CYC7* encodes the minor iso-2-cytochrome c [Sherman and Stewart, 1971].

It is clear that the yeast cell is poised to express the entire cytochrome chain in a coordinated manner. Expression is highest when cells grow in media with a nonfermentable carbon source, under which conditions all ATP synthesis derives from electron transport. Expression of all the cytochromes is turned down five- to tenfold in media containing glucose as carbon source. Cytochrome expression appears to be quite low under anaerobic conditions, but quantitation is difficult because the cytochrome cofactor heme is not made in significant quantities in the absence of oxygen. Several questions can be posed from these observations. How is expression of nuclear genes encoding cytochromes regulated? What coordination exists between this gene family and the mitochondrial cytochrome genes? Finally, what relationship, if any, exists between synthesis of cytochromes and that of related proteins or factors, such as other hemoproteins or heme itself?

II. *CYC1*

As an entry to this formidable problem, we have initiated studies on the nuclear genes encoding cytochromes c with emphasis on *CYC1*. We have attempted to identify the physiological signals, sequences in *cis* to the gene, and *trans*-acting proteins that are involved in *CYC1* control. There is hope that these findings can then be related to *CYC7* and ultimately to genes encoding other cytochromes in the yeast cell.

Cytochrome c is synthesized in the cytoplasm and targets to the mitochondria without possessing any apparent signal sequence [Zimmerman et al., 1981]. The first step in import appears to be association of the apocytochrome with a receptor on the mitochondrial surface followed by heme attachment as the protein is internalized to the outer face of the inner membrane (Fig. 2) [Hennig and Newport, 1981]. Unlike the other cytochromes, c is only peripherally associated with the inner membrane.

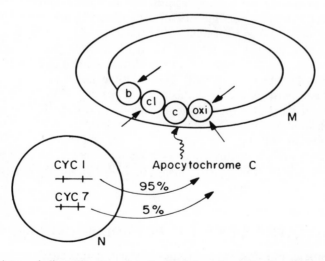

Fig. 2.   Arrows indicate the sites of synthesis of the polypeptides for cytochromes b, cl, c, and c oxidase (oxi). As described in the text, nuclear genes encode the apoproteins of cytochromes c and cl as well as four of the seven subunits of cytochrome c oxidase; the genes for apocytochrome b and for the remaining three subunits of cytochrome c oxidase are mitochondrial.

Many mutations have been identified by classical means, which reduce expression of *CYC1*. Several hundred base substitution mutations at the *CYC1* locus lie in coding sequence [Sherman et al., 1974]. Curiously absent from this collection are possible promoter mutations. Mutations unlinked to *CYC1* were also uncovered. *CYC2* and *CYC3* appear to be involved in heme attachment and/or import of the cytochrome, *CYC4* affects biosynthesis of the cofactor heme, and *CYC5* and *CYC6* exert small effects—their role in determining cytochrome c levels is not clear [Sherman et al., 1965; Matner and Sherman, 1982]. Thus no clear examples of regulatory loci governing *CYC1* transcription emerged from this exhaustive genetic analysis. Mutant hunts based on cytochrome c assays thus cast a large net in which many mutations affecting posttranslational modification, import, or cofactor biosynthesis or attachment would be caught.

To simplify the task of studying *CYC1* regulation, we initiated our studies employing a *CYC1-lacZ* fusion in which the *lacZ* product, β-galactosidase, is a measure of expression and regulation of the *CYC1* promoter [Guarente and Ptashne, 1981]. The fusion bears only the initiator ATG from *CYC1* to avoid complications of posttranslational modification and/or import of the hybrid protein. The fusion can be introduced into yeast in any of three ways. First is by an autonomously replicating plasmid that bears the origin of replication from the yeast 2 μm circle [Beggs, 1978]. These vectors are relatively stable in yeast (lost at 1–10% in overnight culture) and exist in a copy number of about 20 per cell. Second is a plasmid bearing the autonomously replicating sequence, ARS1, and the centromere from chromosome 4, CEN4 [Struhl et al., 1979; Stinchcomb et al., 1982]. This vector is about as stable as the 2 μm plasmid and exists at a copy number of one per cell. Third is integration of the fusion at the *CYC1* locus by homologous recombination between an originless plasmid and the yeast chromosome 10 [Hinnen et al., 1978]. The integrants exist in one copy per cell and are very stable (lost at <0.5% of the cells in an overnight culture). All plasmids used above also bear markers allowing their selection in yeast ($URA3^+$ or $LEU2^+$) and *Escherichia coli* (Amp^R) and the colE1 origin.

III. REGULATION OF *CYC1* BY PHYSIOLOGICAL SIGNALS

Cytochrome c, like other cytochromes in yeast, is regulated by the carbon source in the growth media. Levels are repressed about five- to tenfold in glucose compared to a nonfermentable carbon source such as lactate [Sherman and Stewart, 1971]. In the case of *CYC1*, glucose results in a similar reduction in mRNA levels [Zitomer et al., 1979]. Like the electron transport chain, the TCA cycle, also housed within the mitochondria, is glucose-repressed [Hoosein and Lewin, 1984]. Teleologically, repression of these functions is reasonable; their requirement in ATP production is minimized when cells grow in glucose. Whether the mechanism of glucose repression of mitochondrial functions bears any similarity to that of other systems, such as those involved in maltose or sucrose utilization, remains to be seen.

In addition to carbon source effects, anoxia or anaerobic growth appears to have an inhibitory effect on levels of electron transport cytochromes [Ephrussi and Slonimski, 1950]. Interpretation of this effect is complicated by the fact that synthesis of the cofactor heme requires molecular oxygen at a late step in the pathway [Mattoon et al., 1979]. Thus any reduction in levels of apocytochrome c could be owing to either an instability of the apoprotein or a failure to synthesize the protein under anaerobiosis.

An ambiguity introduced by possible instability of the apocytochrome was removed by use of a *CYC1-lacZ* fused gene. β-Galactosidase, whose levels were under control of *CYC1* regulatory signals, was reduced in amount 200-fold by growth under anaerobic conditions [Guarente and Mason, 1983]. The effect of anaerobiosis on *CYC1* expression was shown to be due to heme deficiency per se. A mutant blocked in the first step of heme biosynthesis also failed to express β-galactosidase unless a heme supplement was added to the media. Direct mRNA determinations showed that *CYC1* message was absent under conditions of heme deficiency [Guarente and Mason, 1983]. Further, by employing heterologous regulatory signals, *CYC1-lacZ* mRNA and β-galactosidase could be synthesized under heme-deficient conditions. This means that the effect of heme on the intact *CYC1* promoter is at the level of initiation of transcription.

The fusion, as expected, was derepressed about tenfold by shift from glucose to lactate under heme-sufficient conditions. *CYC1* is thus regulated over a 2,000-fold range by heme and by carbon source. Carbon source derepression could be attributed, in part, to an increase in intracellular heme levels. The addition of the heme analog deuteroporphyrin IX to glucose media resulted in a threefold derepression of the fusion [Guarente et al., 1984]. Full derepression thus required a component other than heme.

In that heme appears to be a key regulatory molecule in yeast, it is important to consider what physiological signals regulate levels of heme in cells. First, as mentioned above, synthesis of heme is limited by anoxia. Second, levels of several enzymes involved in heme biosynthesis are re-pressed by glucose. *HEM2*, in particular, the gene encoding the second step in the pathway, appears to be highly regulated [Mahler and Lin, 1978]. Third, certain enzymes of the pathway are imported from the cytoplasm into the inner space of the mitochondria (Fig. 2). This is true for the first step (δ-amino levulinate synthetase) and late steps of the pathway. Regulation of heme biosynthesis could thus occur at the level of import of these enzymes or perhaps exit of heme from the mitochondria to the cytoplasm. It is presumed that specific regulatory proteins could bind to heme in the cytoplasm and migrate to the nucleus to regulate transcription. Subsequent sections will discuss experiments designed to elucidate sites *in cis* to *CYC1* and regulatory genes *in trans* that appear to be involved in *CYC1* control.

IV. REGULATORY SITES IN *CIS* TO *CYC1* AND UPSTREAM ACTIVATION

In prokaryotic systems, base substitution mutations obtained by standard genetic means have been extremely useful in identifying promoter and regulatory regions. In the case of the *CYC1* gene, of hundreds of $cyc1^{-1}$ point mutations mapped to the *CYC1* locus, none were outside of the coding sequence [Sherman et al., 1974]. Gene fusion experiments clearly indicated that the promoter-regulatory region of *CYC1* lay upstream of coding DNA. From what follows below, it will become clear that this paradox is due to the redundancy of all the *CYC1* regulatory sites.

S1 mapping of *CYC1* transcripts indicated that initiation occurred at seven or more sites within a 35 base pair region between 50 and 100 nucleotides upstream of coding DNA [Guarente and Mason, 1983; Faye et al., 1981]. Deletion analysis indicated that the three TATA box sequences lying between 90 and 120 nucleotides upstream of the coding sequence were required for efficient transcription (Fig. 3) [Guarente and Mason, 1983]. Also highlighted

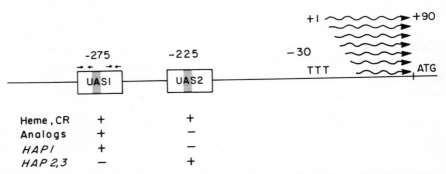

Fig. 3. Locations of upstream activating sequences associated with the coding sequence for apocytochrome cl and a table summarizing activation or inactivation by various agents including *HAP* gene products.

by the deletions was a region 225–300 nucleotides upstream from the region of transcription initiation. Deletion of this upstream activation site (UAS) abolished *CYC1* transcription (Fig. 3) [Guarente and Mason, 1983]. Deletion or spacer mutations of 100–300 base pairs between the UAS and TATA regions did not abolish upstream activation, although variable quantitative effects were observed [Guarente and Hoar, 1984].

To probe whether or not the UAS or TATA regions mediated regulation by heme and by carbon source, the method of promoter fusion was employed. The utility of this approach had been established by an earlier experiment in which the UAS of the *GAL10* gene (galactose-inducible) was substituted in place of the *CYC1* UAS [Guarente et al., 1982]. This promoter fusion gave rise to transcripts that initiated at the normal *CYC1* sites but were regulated by galactose [Guarente and Mason, 1983]. To apply this method to the *CYC1* UAS, DNA fragments encompassing all or portions of the *CYC1* UAS region were placed upstream of the *LEU2* gene TATA box [Guarente et al., 1984]. The latter gene is not normally regulated by heme or carbon source. The constructions also removed *LEU2* upstream sequences required for normal *LEU2* expression. The hybrid promoter bearing the *CYC1* UAS was active and responded to heme and carbon source in much the same manner as *CYC1*. Moreover, two distinct fragments of the *CYC1* upstream region were found to possess UAS activity, one at about −275 (UAS1) and the other at −225 (UAS2) in the intact *CYC1* promoter. Each site responded to both heme and carbon source, and the two sites shared a nine nucleotide core of homology (Fig. 3). The UAS1 site was shown to be responsible for most of the basal expression in glucose, and both sites functioned equally under derepressed conditions. The derepression was thus five-to tenfold for UAS1 and 100-fold for UAS2. Both sites were completely inactive in the absence of heme.

The five- to tenfold derepression of UAS1 in derepressing media could also be elicited by the addition of heme analogs to glucose-grown cells [Guarente et al., 1984]. The analogs had no effect on levels of expression of UAS2 in glucose. Thus it was concluded that derepression of UAS1 was probably due to an increase in intracellular heme levels under derepressed conditions, whereas derepression of UAS2 occurred by some other mechanism. This explained the earlier results in which the intact *CYC1* promoter was only partly derepressed by the heme analog in glucose. These results suggested that UAS1 and UAS2 are regulated by distinct activator proteins.

A more detailed analysis of UAS1 has been initiated employing an in vitro site-directed mutagenesis technique to be described elsewhere (Lalonde and Guarente, unpublished data). The method involved use of α-S dNTPS to generate single base substitutions, which may be transitions or transversions. Mutagenized DNA is transformed into yeast, and colonies with a reduced level of expression of a *CYC1-lacZ* fusion are identified on plates containing the β-galactosidase chromogenic substrate XG. Single base changes resulting in a two-to tenfold reduction in UAS1 activity have been isolated. The mutations highlight a pentameric sequence that is repeated on either side of the UAS1 core sequence (Fig. 3). Thus we surmise that a *trans*-acting activator protein interacts with this pentamer, whereas the role of the core is unclear at present.

UASs have also been found for other yeast genes, such as *GAL10* [Guarente et al., 1982], *HIS3* [Struhl, 1982], and *HIS4* [Donahue et al., 1983]. Indeed, it is likely that regulation of yeast genes, generally, will be governed by UASs. The flexibility allowed in the spacing between UASs and TATA boxes permits a strategy in which tandem multiple UASs could mediate activation of the same transcript in response to different physiological signals. A major outstanding question is how UASs transmit their effects over hundreds of base pairs. Do the sites, for example, allow entry to the template of RNA polymerase or some other factor required for transcription initiation, or do UASs propagate a structural change over several hundreds of base pairs of chromatin that renders the TATA box accessible? One result arguing against the latter hypothesis is the finding that the *CYC1* UAS and *GAL10* UAS do not activate transcription when they are placed to the 3' side of the initiation region [Guarente and Hoar, 1984; Struhl, 1984]. This result contrasts to findings on mammalian enhancers, which do function when positioned downstream of the initiation region and, in some cases, naturally occur in such a location [Banerji et al., 1981; Moreau et al., 1981; Wasylyk et al., 1983]. Also, unlike enhancers, the UASs are ineffective when separated by more than several hundred base pairs from a downstream TATA box [Guarente and Hoar, 1984]. Thus the results to date favor the hypothesis that UASs are entry sites for transcription factors. In that the *CYC1* UASs function when inverted with respect to the TATA box [Guarente and Hoar, 1984], such factors would be able to diffuse in either direction after entry to the template.

V. REGULATORY LOCI GOVERNING *CYC1* CONTROL

Clearly, a complete understanding of upstream activation in yeast requires that the proteins that specifically interact with the UASs be identified, isolated, and employed in a purified in vitro transcription system. A first step to identifying such regulatory proteins is the isolation of regulatory mutations in the genes that encode them. Loci believed to encode activators for genes involved in galactose catabolism [Johnston and Hopper, 1982; Laughon and Gesteland, 1982] and certain amino acid biosynthetic genes [Hinnebusch and Fink, 1983] have been identified genetically.

Trans-acting mutations that prevented activation of a *CYC1-lacZ* fusion were isolated by employing fusions whose expression was driven *either* by UAS1 or UAS2. Vectors deleted for one site or the other were employed for this purpose. A mutation in one locus, *HAP1*, prevented activation of UAS1 even when the heme analog was supplied in the media [Guarente et al., 1984]. The *hap1-1* mutation did *not* affect the activity of UAS2. Similarly, mutations in a locus *HAP2* reduced the activity of UAS2 without exerting a significant effect on UAS1 (Fig. 3) [Guarente et al., 1984]. The specificity of the *hap* mutations for UAS1 or UAS2 is consistent with the view that the sites are activated by different proteins.

It is possible to surmise what other cytochrome genes are regulated by the *HAP1* and *HAP2* loci by an examination of the pleiotropy of the mutant alleles. In this regard, the *hap1-1* mutation, which completely eliminates the activity of UAS1, has little effect on cellular cytochromes other than c (Mason and Guarente, unpublished data). In contrast, the *hap2-1* mutation causes a reduction in all cytochromes giving rise to a respiration-deficient strain [Guarente et al., 1984]. It is not clear whether the *HAP2* product is a direct activator of all of the cytochrome genes or controls synthesis of one or a series of activators of this gene family. Transcription of mitochondrial cytochrome genes is not affected by *hap1* or *hap2* mutations (Prezant and Guarente, unpublished data). Thus a picture begins to emerge in which nuclear cytochrome genes are coregulated by one system and cytochrome c is specifically activated by a second system. The two *trans*-acting factors can act independently, because the UAS1 and UAS2 targets are discrete sites. At present, there is no evidence that suggests tight coupling between nuclear and mitochondrial cytochrome gene transcription.

VI. OTHER CYTOCHROME GENES

A long-term goal of the study of cytochrome gene control is understanding how coregulation between genes encoding different cytochromes is achieved. An initial approach described above is to examine the effects of *trans*-acting mutations on levels of various cellular cytochrome genes. To obtain more detailed information, it is necessary to isolate clones of the various cytochrome genes. The clones enable quantitative determinations of mRNA levels under various physiological conditions and in various *hap* mutant strains.

Further, by constructing fusions to *lacZ* followed by in vitro mutagenesis, it is possible to identify sequences in cis that mediate regulation. A comparison of these sequences and the relevant trans-acting loci will enable a precise molecular framework of gene coordination to be set up. This sort of approach has proved most useful for amino acid biosynthetic genes under the general control system [Donahue et al., 1983].

The gene encoding the minor iso-2-cytochrome c, *CYC7*, has been cloned. By employing this clone, we have constructed *lacZ* fusions and determined that *CYC7* expression appears to be affected by mutations in *HAP1* (Prezant and Guarente, unpublished data). Thus we might predict that *CYC7* would contain a UAS site that would bear the UAS1 pentameric sequence and require the action of *HAP1* to be activated.

Like *CYC1*, *CYC7* is under catabolite repression control. Surprizingly, expression of the gene is not reduced by heme deficiency [Laz et al., 1984]. This means that the *HAP1* product might not require heme for its synthesis or activity. It is possible that the regulation of *CYC1* by heme is mediated by a regulatory system independent of *HAP1* or *HAP2*. For example, it is conceivable that a repressor could bind to UAS1 and UAS2 in the absence of heme to prevent activation by the *HAP* systems. The region of homology between the UASs would be a reasonable candidate for such an operator site. Alternatively, a repressor could inactivate the *HAP* product by a protein-protein interaction, which is believed to occur in the case of proteins regulating the genes of the galactose utilization system in yeast [Oshima, 1982]. A distinct view is that control is entirely positive and that heme increases the levels of *HAP* products or alters their specificity by binding to them to allow activation of *CYC1*. Further genetic and molecular analysis will be required to distinguish between these various possibilities.

Experiments are also in progress to clone other nuclear genes encoding cytochromes, such as cytochromes c1 and oxidase subunits. The availability of these clones will enable the execution of experiments similar to those performed on *CYC1* and *CYC7*. Further, the clones might be used to make mutations in these genes to study the effects of deficiency in one cytochrome on the regulation of others.

VII. REGULATION OF HEME BIOSYNTHESIS

It is of interest to determine the extent to which heme biosynthesis is coupled to cytochrome gene expression. To this end, we have isolated mutations in six of the steps of the heme biosynthetic pathway, determined which enzyme was affected by each lesion, and cloned the corresponding wild type genes from a yeast library by complementation of the heme auxotrophy. We envision two possible points of regulation of the pathway. First is transcriptional control, which might account for the observed regulation by catabolite repression of the first two enzymes in the pathway [Mahler and Lin, 1978]. Second is control of import into the mitochondria of enzymes catalyzing the first (δ-amino levulinate synthetase) or late steps of the pathway (Fig. 1). Enzymes in the middle portion of the biosynthetic pathway are located in the cytoplasm.

To probe for regulation of transcription or of import by physiological signals (heme levels or catabolite repression) or by *HAP* loci, we have constructed fusions of *lacZ* to genes encoding the first, second (δ-amino levulinate dehydratase), and last (ferrochelatase) enzymes of the pathway. By determining levels and intracellular location (cytoplasm or mitochondria) of the β-galactosidase fusion proteins, we hope to deduce the magnitude and nature of regulatory strategies employed to regulate heme biosynthesis in yeast.

ACKNOWLEDGMENTS

This work was carried out at the Department of Biology, M.I.T., and was funded by N.I.H. grant 5-R01-GM30454 to L.G.

VIII. REFERENCES

Banerji J, Rusconi S, Schaffner W (1981): Cell 27:299.

Beggs J (1978): Nature 275:104–109.

Donahue T, Daves R, Lucchini G, Fink G (1983): Cell 32:89–98.

Ephrussi B Slonimski P (1950): Biochim Biophys Acta 6:256–257.

Faye G, Leung D, Tatchell K, Hall B, Smith B (1981): Proc Natl Acad Sci USA 78:2258–2262.

Guarente L, Hoar E (1984): Proc Natl Acad Sci USA (in press).

Guarente L, Lalonde B, Gifford P, Alani E (1984): Cell 36:503–511.

Guarente L, Mason T (1983): Cell 32:1279–1286.

Guarente L, Ptashne M (1981): Proc Natl Acad Sci USA 78:2199–2203.

Guarente L, Yocum R, Gifford P (1982): Proc Natl Acad Sci USA 79:7410–7114.

Hennig B, Newport W (1981): Eur J Biochem 121:203–212.

Hinnebusch A, Fink G (1983): Proc Natl Acad Sci USA 80:5374–5378.

Hinnen A, Hicks J, Fink G (1978): Proc Natl Acad Sci USA 75:1929–1933.

Hoosein M, Lewin A (1984): Mol Cell Biol 4:247–253.

Johnston S, Hopper J (1982): Proc Natl Acad Sci USA 79:6971–6975.

Katan M, Pool L, Groot G (1976): Eur J Biochem 65:95.

Laughon A, Gesteland R (1982): Proc Natl Acad Sci USA 79:6827–6831.

Laz T, Pietras D, Sherman F (1984): Proc Natl Acad Sci USA 81:4475–4479.

68   Guarente

Lin L, Beattie D (1976): In Bücher T, et al. (eds); "Genetics and Biogenesis of Chloroplasts and Mitochondria." p 281.

Mahler H, Lin C (1973): J Bacteriol 135:54–61.

Mason T, Schatz G (1973): J Biol Chem 248:1355.

Matner R, Sherman F (1982): J Biol Chem 257:9811–9821.

Mattoon J, Lancashire W, Sanders H, Carvajal E, Malamud D, Bray G, Panek A (1979): In Biochemical and Clinical Aspects of Oxygen." pp 421–435.

Moreau P, Hen R, Wasylyk B, Everett R, Gaub M, Chambon P (1981): Nucleic Acids Res 9:6047–6069.

Oshima Y (1982): In Strathern JN, et al. (eds); "The Molecular Biology of the Yeast Saccharomyces, Metabolism and Gene Expression." Cold Spring Harbor, NY: Cold Spring Harbor Laboratory, pp 159–180.

Sherman F, Stewart J (1971): Annu Rev Genet 5:257–296.

Sherman F, Stewart J, Jackson M, Gilmore R, Parker J (1974): Genetics 77:255–284.

Sherman F, Taber H, Campbell W (1965): J Mol Biol 13:21–39.

Stinchcomb D, Mann C, Davis R (1982): J Mol Biol 158:157–179.

Struhl K (1982): Nature 300:284.

Struhl K (1984): Proc Natl Acad Sci USA (in press).

Struhl K, Stinchomb D, Scherer S, Davis R (1979): Proc Natl Acad Sci USA 76:1035–1039.

Wasylyk B, Wasylyk C, Augereau P, Chambon P (1983): Cell 32:503–514.

Zimmerman R, Hennig B, Weuport W (1981): Eur J Biochem 116:455–460.

Zitomer R, Montgomery D, Nichols D, Hall B (1979): Proc Natl Acad Sci USA 76:3627–3631.

Molecular Developmental Biology, pages 69–82
© 1986 Alan R. Liss, Inc.

Molecular Structure of Telomeres in Lower Eukaryotes

Elizabeth H. Blackburn

Department of Molecular Biology, University of California at Berkeley, Berkeley, California 94720

I. INTRODUCTION

Telomeres are the ends of linear eukaryotic chromosomes. Many cytogenetic observations have been made that argue that they must have highly specialized properties [reviewed in Blackburn and Szostak, 1984]. One functional aspect of chromosomal telomeres is most apparent when their behavior in vivo is contrasted with that of newly broken chromosome ends, produced by either mechanical rupture or X-irradiation. Such freshly broken ends generally will undergo fusion with other broken ends, or are highly recombinogenic, and in some situations are subject to degradation. In contrast, normal telomeres do not show this reactivity, suggesting that one role of a telomere is to protect and stabilize the end of the chromosome.

Another question that needs to be addressed when the structure of telomeres is considered is that of their replication. All known DNA polymerases require a primer (either DNA or RNA) bearing a 3'-hydroxyl group to initiate DNA synthesis. In the case of an RNA primer, it is removed once synthesis has been primed. The removal of the primer would result in a 5'-terminal gap being left in one strand at each end of a replicated linear DNA molecule. Although linear DNA viruses have solved this problem in a variety of ways, there was until recently no direct evidence on how the ends of linear eukaryotic chromosomes are able to be maintained indefinitely through repeated rounds of replication. Various models were proposed to explain telomere replication [reviewed in Blackburn and Szostak, 1984]. However, structural analysis of telomeres was necessary before such models could be evaluated.

II. LOWER EUKARYOTIC SYSTEMS FOR STUDYING TELOMERE STRUCTURE

The lengths of eukaryotic chromosomes commonly exceed thousands of kilobase pairs (kb). Direct analysis of such chromosomal telomeres was therefore impracticable until, as described below, recent methods for cloning at least some telomeres in yeast were developed. Molecular and biochemical analyses of chromosomal termini were confined to systems with naturally occurring, short, linear nuclear DNAs.

A. Ciliated Protozoa

Ciliated protozoa are useful systems for studying the molecular structures of telomeres. Each single-celled organism contains two types of nuclei: the diploid, germline micronucleus and the polygenomic, somatic macronucleus, which is the site of gene expression and which divides amitotically in vegetative growth. The micronucleus divides mitotically during vegetative growth and carries out meiosis in the sexual process, conjugation. After conjugation, each cell receives a new macronucleus differentiated from a mitotic sister of the micronucleus, and the old macronucleus is destroyed. Several studies have shown that a profound reorganization of the genome occurs in macronuclear differentiation in many ciliates. Although the quantitative and cytological properties of this genomic reorganization process are very different among the different classes of ciliates, examination of their resulting macronuclear genomes has revealed major common features even between organisms as apparently unlike in their macronuclear development as *Tetrahymena* (a holotrichous ciliate) and *Oxytricha* (a hypotrichous ciliate) [reviewed in Blackburn et al., 1983].

One aspect of the genomic reorganization process important for studies on telomere structure is that the chromosomes comprising the micronuclear genome are converted by a highly specific process into defined subchromosomal DNA molecules. These macronuclear DNAs are linear and therefore require telomeric sequences at their termini in order to be stably maintained and replicated. Other rearrangements involving deletions of specific DNA sequences, and ribosomal RNA gene (rDNA) amplification, also occur in macronuclear development [reviewed in Blackburn et al., 1983, 1985].

In the holotrichous ciliate *Tetrahymena thermophila,* rearrangement and amplification of the single, chromosomally integrated copy of the ribosomal RNA gene results in its conversion to several thousand linear rDNA molecules in the macronucleus. These relatively short (21 kb) molecules are in the form of palindromic dimers in the mature macronucleus, each consisting of two rRNA genes in head-to-head arrangement. Their small size and relative abundance allowed direct analysis of their telomeric sequences and structures [Blackburn and Gall, 1978]. Early in macronuclear development, free rRNA genes are also found in the form of linear, self-replicating single genes [Pan and Blackburn, 1981], although these are lost from mature macronuclei. A similar molecular form of the rDNA is found in the mature macronucleus of the related tetrahymenid *Glaucoma chattoni* (Fig. 1) [Katzen et al., 1981]. Small linear DNAs (0.5–20 kb in length) comprise the entire DNA content of macronuclei of several hypotrichs [Prescott and Murti, 1974], and advantage has been taken of this fact to analyze their telomeric

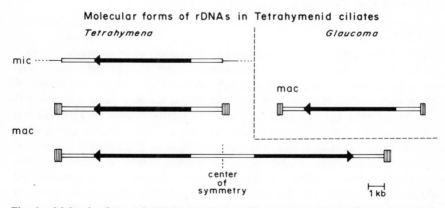

Fig. 1. Molecular forms of rRNA genes in the ciliated protozoans *Tetrahymena* and *Glaucoma*. A single chromosomally integrated rRNA gene, indicated by the bar, is found in the micronuclear genome of *T. thermophila* flanked by micronuclear-limited DNA sequences as indicated by horizontal lines [King and Yao, 1982]. The comparable sequence in *Glaucoma* has not been analyzed. Solid bars and arrowheads indicate rRNA transcription units and $5' \rightarrow 3'$ polarity of transcripts; open bars are nontranscribed spacer regions. Vertical bars are telomeric CCCCAA repeats [Blackburn and Gall, 1978; Katzen et al., 1981; Challoner and Blackburn, unpublished results]. Adapted from Blackburn et al. [1985], with permission.

structures [Herrick and Wesley, 1978; Oka et al., 1980; Klobutcher et al., 1981].

B. Slime Mold rDNAs

The extrachromosomal rDNAs of the slime molds *Physarum* and *Dictyostelium* are, like those of *Tetrahymena,* linear, palindromic molecules, 61 and 87 kb in length, respectively. They are amplified to high copy number relative to chromosomal DNA sequences, and this, combined with their relatively short lengths, has aided analysis of their telomeric structures [Johnson, 1980; Emery and Weiner, 1981].

C. Trypanosomes

In at least one trypanosome species (e.g., *Trypanosoma brucei*), which can live in the bloodstream of its mammalian host, a large fraction of the nuclear DNA is organized in the form of minichromosomes bearing genes for variant surface glycoproteins [Williams et al., 1982; Van der Ploeg et al., 1984]. Their sizes are in the range of $\sim 10^2$ kb. It has been estimated that there are $\sim 10^2$ such minichromosomes per nucleus in *T. brucei* [Van der Ploeg et al., 1984]. The resulting abundance of chromosomal telomeres in this organism made possible direct labeling and sequence analysis of telomeric sequences [Blackburn and Challoner, 1984].

D. Yeast

One of the more favorable systems for direct analysis of telomere structure and function has been the yeast *Saccaromyces cerevisiae,* because this organism can be transformed by exogenously added DNA. To test the ability of *Tetrahymena* DNA termini to function in yeast, a linear yeast plasmid was constructed in vitro by ligating the purified terminal regions of *Tetrahymena* rDNA to a yeast vector [Szostak and Blackburn, 1982]. In this approach, a circular yeast plasmid vector, carrying a selectable marker and capable of replicating autonomously, was linearized with a single restriction enzyme cut. In vitro ligation of the terminal restriction fragments of the *Tetrahymena* rDNA molecules described above to each end of this plasmid allowed it to be maintained and replicated in yeast as a free linear molecule. Removal of one end of this "linear vector" prevented it from being stably maintained in yeast. Thus shotgun cloning of yeast genomic DNA restriction fragments onto the cut end of the linear vector was done to select for yeast chromosomal fragments that conferred the ability to replicate as a free linear molecule. These yeast fragments were identified as genomic telomeres [Blackburn and Szostak, 1982; Shampay et al., 1984].

The short linear plasmids so constructed were maintained in multiple copies per cell, facilitating structural analysis of their terminal or telomeric structures. Furthermore, this system provides a way of assaying for telomere function, as defined by the ability of a telomere to stabilize and allow continued replication of a linear DNA molecule in the nucleus.

III. TELOMERIC DNA SEQUENCES AND STRUCTURES
A. DNA Sequences of Telomeres

The DNA sequences at the extreme ends of the chromosomal and other linear DNA molecules in the lower eukaryotic systems described above have been determined. They are simple and satellitelike, consisting of short G + C-rich tandem repeats. It is likely that these simple telomeric sequences are the essential functional components of a telomere, their presence being sufficient to stabilize a chromosomal end.

The first such sequence defined was that of the amplified rDNA molecules of the macronucleus of *Tetrahymena thermophila* [Blackburn and Gall, 1978]. Each end of the rDNA molecule was found to consist of over 50 tandem repeats of the hexanucleotide CCCCAA [Blackburn et al., 1983]. Several single-strand, one-nucleotide gaps were found in the C_4A_2 strand at specific positions within the cluster of repeats. On the G_4T_2 strand, at least one single-strand gap was found at a position inferred to be internal to the C_4A_2 strand breaks. The single-strand breaks were confined to the terminal (distal) ~ 100 bp region of this block of repeats [Blackburn et al., 1983]. The structure of the rDNA telomeres inferred from these studies, and from results described below, is shown in Figure 2.

Similar repeats comprise the telomeres of the other linear macronuclear DNAs of *Tetrahymena* [Yao and Yao, 1981] and the related holotrichous ciliate *Glaucoma* [Katzen et al., 1981], as first shown by Southern blotting and direct sequence analysis, respectively. Sequence comparison of several such cloned macronuclear rDNA and non-rDNA telomeres of *Tetrahymena* has shown that the repeated CCCCAA sequence is the only sequence shared between different macronuclear DNAs [Blackburn et al., 1985; Challoner, Ryan, Cherry, and Spangler, unpublished observations).

In a different class of ciliates, the hypotrichs, such as *Stylonychia* and *Oxytricha,* the macronuclear DNA telomeric sequence consists of tandem repeats of CCCCAAAA, and the structure of the termini of purified macronuclear DNAs can be written [Oka et al., 1980; Klobutcher et al., 1981]:

$$5'\ C_4\ A_4\ C_4\ A_4\ C_4\ \ldots$$
$$3'\ G_4\ T_4\ G_4\ T_4\ G_4\ T_4\ G_4\ T_4\ G_4\ \ldots$$

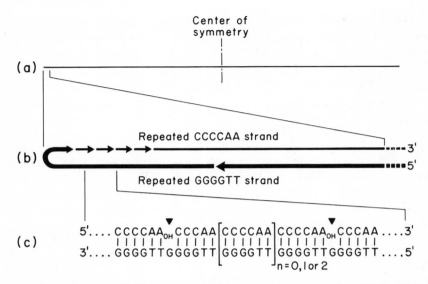

Fig. 2. Structure of the terminal regions of the palindromic macronuclear rDNA molecule of *Tetrahymena thermophila* [Blackburn and Gall, 1978; Blackburn et al., 1983]. a, The palindromic rDNA; b, the terminal few hundred base pairs. Positions of specific single-strand breaks on the repeated GGGGTT strand (thick line) and repeated CCCCAA strand (medium-thick line) are shown as arrowheads. The innermost break is ~ 100 base pairs from the extreme end of the molecule, which is shown here as a fold-back structure formed from the GGGGTT strand (see text). Arrows indicate $5' \rightarrow 3'$ polarity of each strand. c, DNA sequence of the region encompassing two adjacent single-strand breaks (▼) on the CCCCAA strand. Adapted from Blackburn [1982], with permission.

The end of the macronuclear DNA is on the left side. However, it is not known whether the $3'$ termini of the DNA molecules in vivo are single-stranded as shown; there might be base-paired DNA and/or RNA segments attached to the extreme termini of these molecules that are lost during extraction and purification of macronuclear DNA for sequence analysis [Dawson and Herrick, 1982].

The termini of the rDNA molecules of *Physarum* behave in a manner very similar to those of *Tetrahymena* when subjected to a variety of in vitro labeling protocols (suggesting that they too have single-strand breaks near their ends). Evidence for tandem repeats of the sequence $CCCTA_n$ was found [Johnson, 1980]. The termini of the *Dictyostelium* rDNA molecules contain a somewhat different simple sequence: an irregular, satellitelike DNA sequence of tandem repeats of the general formula $C_{1-8}T$ [Emery and Weiner, 1981]. Using in vitro labeling protocols like those first used to identify the telomeric sequences of the rDNA of *Tetrahymena*, the sequence CCCTAA was found as tandem repeats at the termini of the chromosomes of the hemoflagellate *Trypanosoma brucei* [Blackburn and Challoner, 1984]. These repeats occur in restriction fragments of *T. brucei* genomic DNA identified as telomeric on the basis of their sensitivity to BAL31 nuclease [Blackburn and Challoner, 1984; Van der Ploeg et al., 1984]. In yeast, the most distal ~ 500 bp of a chromosomal telomere cloned on a linear plasmid vector as described above consists entirely of tandem repeats of the irregular repeated sequence $C_{1-3}A$ [Shampay et al., 1984]. Like *Tetrahymena* rDNA molecules, the yeast telomere has specific single-strand breaks in the distal ~ 100 bp of the molecule, allowing nick translation of this region by *Escherichia coli* DNA polymerase I. Southern blot analysis of genomic yeast DNA has shown that $C_{1-3}A$ repeats are likely to be common to all yeast chromosomal ends, although the adjacent chromosomal sequences can vary [Shampay et al., 1984]. The seemingly irregular $C_{1-3}A$ repeat units that comprise the yeast telomere, although reminiscent of the irregular $C_{1-8}T$ tandem repeats at the termini of *Dictyostelium* rDNA molecules, nevertheless all conform to a larger repeat unit, which can be written $5'[C_{2-3}A(CA)_{1-3}]3'$. This finding suggests that there are strong constraints, imposed during either their formation or maintenance, on the arrangement of the basic $C_{1-3}A$ repeats in yeast.

The observation that C_4A_2 repeats at the end of a linear DNA could stabilize such a molecule in yeast has been further explored by sequence analysis of the terminal portion of the *Tetrahymena* rDNA-derived end of a linear plasmid in yeast. The presence in the cloned yeast telomere of single-strand breaks within a sequence very rich in C and A residues on one strand had been suggested by the specific incorporation by DNA polymerase I from *E. coli* of radioactive dCTP + dATP or dGTP + dTTP (but not other pairwise combinations of deoxyribonucleoside triphosphates) into a linear yeast plasmid with one *Tetrahymena* rDNA-derived end and a cloned yeast telomere at the other end [Szostak and Blackburn, 1982]. The incorporated label was further localized by restriction digestions to the most terminal fragment of the *Tetrahymena* rDNA end and the most terminal fragment of the yeast telomere end. Pyrimidine tract analysis of these fragments labeled by incorporation of $[\alpha\text{-}^{32}P]$-dCTP and -dATP produced the labeled pyrimidine tracts $C_3(A)$, $C_2(A)$, and $C(A)$ from the yeast end and $C_4(A)$, $C_3(A)$, $C_2(A)$, and $C(A)$ from the *Tetrahymena* rDNA end (Blackburn, unpublished results). Because a similar experiment on *Tetrahymena* rDNA purified from *Tetrahymena* cells produced only $C_4(A)$ tracts [Blackburn and Gall, 1978], this result suggested that *Tetrahymena* rDNA end, although retaining *Tetrahymena*-derived C_4A_2 repeats, also had acquired sequences similar to those at the yeast telomere. By direct DNA sequence analysis, it has been shown that in one clone the *Tetrahymena* rDNA terminus was extended, after maintenance in growing yeast cells, by at least 230 bp of yeast $C_{1-3}A$ telomeric repeats [Shampay et al., 1984]. The arrangements of telomeric sequences in this telomere and in the cloned yeast chromosomal telomere are shown diagramatically in Figure 3.

The tandemly repeated telomeric DNA sequences in all these lower eukaryotic systems are very similar, and can be described by the general formula

$$5'C_{1-8}\overset{A}{(T)}_{1-4}3',$$

in which at least some of the repeats have short runs of C residues. The strand whose sequence is shown here always has a $5' \rightarrow 3'$ polarity from the end toward the interior of the linear DNA molecule or chromosome. Thus telomeric repeats form a pair of inverted sequences, one at each end of any given linear or chromosomal DNA molecule.

B. Accessibility of Telomeric Ends to End-Labeling Reactions

The presence of single-strand breaks in the distal portion of the block of telomeric repeats appears widespread, the possible exception being the telomeres of the macronuclear DNAs of hypotrichous ciliates. However, another shared feature of these evolutionarily diverse telomeres is the apparent inaccessibility of their extreme ends to certain enzymatic end-labeling reactions. To characterize this property further, the native telomeres of the palindromic *Tetrahymena* rDNA molecules shown in Figure 2 and the linear yeast plasmid shown in Figure 3 were analyzed by end labeling. Two reactions were tested: 1) addition of the chain-terminating deoxynucleotide analogue cordycepin to the 3' end of a DNA by terminal deoxynucleotidyl

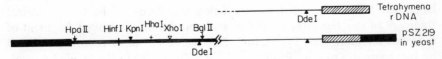

Fig. 3.   Arrangement of telomeric and telomere-associated sequences in a linear yeast plasmid. The linear plasmid pSZ219 has one yeast chromosomal telomere and one *Tetrahymena* rDNA-derived telomere [Shampay et al., 1984]. The top line shows one terminal region of the rDNA from *Tetrahymena* macronuclei, and its relationship to the linear yeast plasmid is shown below. (——), *Tetrahymena* rDNA sequence; ▨, CCCCAA repeats; ▤, yeast chromosomal telomere-associated sequence; ■, $C_{1-3}A$ repeats. Some restriction enzyme cutting sites are indicated.

transferase and 2) exchange of a 5' phosphate of a DNA strand with the $\gamma^{32}P$ of rATP, catalyzed by T_4 polynucleotide kinase.

Terminal transferase acts most efficiently on single-stranded DNA, or single-stranded protruding 3' extensions of a double-stranded DNA [Bollum, 1974]. Divalent cation is required for the reaction: Co^{2+} stimulates addition of nucleotides to a 3'-hydroxyl end by terminal transferase compared with Mg^{2+} for both single-strand 3' extensions and recessed 3' ends of DNA [Roychoudhury et al., 1976]. In contrast, polynucleotide kinase acts more efficiently on a protruding 5' terminus of a DNA strand than a recessed 5' end (i.e., an end of a DNA with a 3' extension) [Lillehaug et al., 1976].

The accessibility of native telomeric ends to these reactions was measured by comparing the amount of label incorporated at telomere ends with various types of restriction-cut ends. *Tetrahymena* rDNA and the linear yeast plasmid shown in Figure 3 were purified, respectively, from *Tetrahymena* and yeast cells. They were cut with restriction enzymes that leave 3' or 5' overhanging ends: HhaI (2 base 3' overhang), PstI (4 base 3' overhang), SacI (4 base 3' overhang), ClaI (2 base 5' overhang), and BamHI (4 base 5' overhang). The various restriction fragments, terminating with the telomere and one of these restriction-cut ends, were labeled in vitro with $[\alpha^{32}P]$cordycepin triphosphate by terminal transferase. This reaction adds a single $[\alpha^{32}P]$cordycepin residue to the 3'-hydroxyl end of a DNA substrate. The effect of using Mg^{2+} or Co^{2+} as the divalent cation was examined by performing parallel reactions in the presence of one or another of these cations. In a similar series of experiments, the restriction fragments were labeled with $[\gamma^{32}P]$rATP by polynucleotide kinase in the phsophate exchange reaction with the 5' end of the DNA substrate [Berkner and Folk, 1977].

To analyze the labeling of the telomeres in these reactions and compare the extents of labeling of telomeres with the restriction enzyme-cut ends, end-labeled fragments were cut with a second restriction enzyme to separate the labeled telomere from the labeled restriction-cut end. The labeled fragments were separated by gel electrophoresis, the gels dried and autoradiographed, and the autoradiograms scanned with a densitometer. The amount of radioactivity in each gel band was quantitated by measuring the area under each peak of the densitometric scans. The ratios of peaks were then calculated (Table I).

The incorporation of labeled nucleotides increased with time over a 30 min period in the case of the terminal transferase reactions and over a 6 min reaction time in the case of the kinase reactions. The results for two time points are shown for each reaction in Table I. For each reaction and each time point, the amount of radioactivity in each peak is shown normalized to

TABLE I. Relative Accessibility of Telomeres and Restriction-Cut Ends of Duplex DNAs to End Labeling

Cation	Time (min)	3' Extension	5' Extension	Telomere

Terminal transferase ([α-^{32}P] cordycepin triphosphate addition to 3' end)[a]
Tetrahymena linear rDNA molecules

					Linear rDNA purified from
Cation	Time (min)	HhaI	PstI	BamHI	*Tetrahymena*
Co^{2+}	10	1.0		0.51	0.32
	30	1.0		0.36	0.22
Co^{2+}	30		1.0		0.46
Mg^{2+}	10	1.0		0.28	2.3
	30	1.0		0.52	1.4

Linear Plasmid in yeast

Cation	Time (min)	SacI	ClaI	Yeast telomere on linear plasmid	*Tet.* rDNA-derived purified from yeast
Co^{2+}	10	1.0	0.58	0.50	0.47
	30	1.0	0.38	0.14	0.10
Mg^{2+}	10	1.0	0.41	1.5	1.7
	30	1.0	0.31	0.62	0.80

5' Polynucleotide kinase ([γ-^{32}P] rATP phosphate exchange with 5' end)[b]
Tetrahymena linear rDNA molecules

Cation	Time (min)	BamHI	Linear rDNA purified from *Tetrahymena*
Mg^{2+}	2	1.0	<0.1
	6	1.0	<0.1

[a]Reactions were carried out at 37°C under conditions similar to those described by Tu and Cohen [1980]. The concentration of telomeric ends was 0.4–1.0 nmolar in these reactions.
[b]Reactions (at 37°C) were similar to those described by Berkner and Folk [1977]. Concentration of telomere ends was ~ 1 nmolar; rADP concentration was 10 μmolar.

1.0 for one labeled restriction-cut end type, and the radioactivity incorporated at the other ends is expressed as a ratio compared to that labeled end.

The results in Table I show two interesting properties of the telomeric ends. First, in the presence of Co^{2+} as the divalent cation in the terminal transferase reactions, the labeling of the telomeres was always less efficient than the labeling of a protruding single-stranded 3' end. However, the labeling of the rDNA telomere by polynucleotide kinase was also less efficient than the labeling of a protruding single-stranded 5' end, despite the fact that there is more than one DNA end, owing to the presence of single-strand breaks, at the telomere (Fig. 2). Similar inaccessibility to end labeling by polynucleotide kinase has been seen with *Dictyostelium* rDNA telomeres [Emery and Weiner, 1981]. Second, the effect of Mg^{2+} on the terminal transferase labeling of telomeres was unexpected; unlike the restriction-cut ends, whose labeling in the presence of Mg^{2+} was decreased relative to the labeling in Co^{2+} as expected, the labeling of telomeres was relatively enhanced. This result suggests that the inaccessibility of the telomere to end labeling is not the result of its being engaged in a conventional Watson-Crick duplex structure; such a DNA duplex would be more stabilized, and hence less accessible to 3' end labeling, by Mg^{2+} than Co^{2+}.

To account for these observations, it is noteworthy that the 3' protruding strand of a telomere would be a G-rich sequence based on the general sequence formula of telomeres described above. Dugaiczyk et al. [1980] have postulated that poly-dG sequences can form two- and three-strand structures mediated by non-Watson-Crick base pairings. Other multistrand structures involving polyG homopolymers have been described [reviewed in Saenger, 1983]. Thus such a structure might exist at the ends of chromosomal DNAs, possibly involving folding back of the protruding G-rich strand upon itself, as depicted in Figure 2 for *Tetrahymena* rDNA. This inaccessibility of telomeric ends conceivably aids in protecting the ends of chromosomal DNAs in the cell from the action of exonucleases or double-strand break-repair systems.

IV. REPLICATION OF TELOMERES

The structural properties of telomeres suggest that they have an unusual mode of replication. Much evidence has accumulated from several lower eukaryote systems to suggest that addition of DNA to telomeres occurs in the course of their replication and might indeed be intrinsic to the replication process [reviewed in Blackburn, 1984]. Telomeres appear to lengthen during their replication over many generations in *Trypanosoma brucei* [Bernards et al., 1983; Pays et al., 1983]. A similar steady lengthening of newly formed, amplified rDNA telomeres has also been seen in *Tetrahymena thermophila*, in the first 40 cell generations following formation of a new macronucleus (Larson and Blackburn, unpublished results). In all systems, however, it appears that the overall length of telomeres is regulated, and indefinite growth of telomeres apparently does not occur. It has been observed in almost every lower eukaryotic system that the telomeric restriction fragments from populations of unsynchronized cells are heterogeneous in length [Blackburn and Gall, 1978; Johnson, 1980; Emery and Weiner, 1981; Blackburn and Challoner, 1984; Shampay et al., 1984; Pays et al., 1983], suggesting that continuous lengthening and shortening of telomeres is occurring in dividing cell populations.

The addition of yeast-specific repeats to *Tetrahymena*-derived C_4A_2 repeats [Shampay et al., 1984], and probably also to C_4A_4 repeats [Pluta et al.,

1984], suggests that the telomeric DNA addition process is not template-directed. A model for telomere replication that takes into account the above observations has been suggested by Shampay et al. [1984]. In this model, a terminal transferaselike activity is responsible for de novo addition of telomeric repeats to the 3' end of the G-rich telomeric sequence. If the preexisting 3' end of this strand is folded or inaccessible as described above, then the addition reaction has to involve the unfolding of this end. The extended single G-rich strand at some stage could become a template for discontinuous synthesis of the complementary C-rich strand by a combination of primase and DNA polymerase reactions. Incomplete ligation of this distal portion of the telomere would then account for the observed single-strand breaks near the end of the DNA.

V. SUMMARY AND FUTURE POSSIBILITIES

Several different lower eukaryotic systems contain short, linear nuclear DNA molecules and have thus been valuable sources of telomeric DNAs. The similarity of the DNA sequences and other molecular features of the telomeres of evolutionarily diverse lower eukaryotes strongly suggests that there is a particular type of DNA structure required for telomere function. Whether or not these similarities will extend to telomeres in multicellular or higher eukaryotes is not yet known.

A remarkable property of telomeres revealed by detailed structural analyses is their ability to gain telomeric DNA sequences by addition to their ends. These findings suggest that a novel form of DNA replication is operating at the ends of chromosomes. It is likely that DNA transformation of yeast, and the wealth of genetic manipulations possible with this organism, will be important in elucidating the mechanism of, and trans-acting factors required for, such a process. In addition, the developmentally controlled formation of new telomeres in the developing somatic nuclei of ciliated protozoans makes them attractive systems for understanding some aspects of telomere growth.

Observations on the molecular dynamics of telomeres might have implications for other aspects of chromosome behavior. In particular, the process off "chromosome healing" has been observed to occur at specific developmental stages in maize and insects [reviewed in Blackburn and Szostak, 1984]. The healed end formed from a broken chromosome acquires the stability of a normal telomeric end. By implication, it acquires telomeric DNA sequences. Whether or not this process shares features in common with the developmentally programmed formation of new telomeres seen in ciliates or nematodes is an intriguing question that awaits analyses at a molecular level.

ACKNOWLEDGMENTS

The author thanks Marcia Budarf, Peter Challoner, Mike Cherry, Carol Greider, Elizabeth Howard, Drena Larson, Thecla Ryan, Beth Spangler, and Janis Shampay for helpful discussions and contributions to this work. This research was supported by grants GM26259 and GM32565 from the National Institutes of Health.

VI. REFERENCES

Berkner K, Folk B (1977): Polynucleotide kinase exchange reaction: Quantitative assay for restriction endonuclease-generated 5'-phosphoryl termini in DNAs. J Biol Chem 252:3176–3184.

Bernards A, Michels PAM, Lincke CR, Borst P (1983): Growth of chromosome ends in multiplying trypanosomes. Nature 303:592–597.

Blackburn EH (1982): Characterization and species differences of rDNA: Protozoans. In Busch H, Rothblum L (eds): "The Cell Nucleus Vol X, Part A." New York: Academic Press, pp 145–170.

Blackburn EH (1984): Telomeres: Do the ends justify the means? Cell 37:7–8.

Blackburn EH, Challoner PB (1984): Identification of a telomeric DNA sequence in Trypanosoma brucei. Cell 36:447–457.

Blackburn EH, Gall JG (1978): A tandemly repeated sequence at the termini of the extrachromosomal ribosomal RNA genes in Tetrahymena. J Mol Biol 120:33–53.

Blackburn EH, Szostak JW (1984): The molecular structure of centromeres and telomeres. Annu Rev Biochem 53:163–194.

Blackburn EH, Budarf M., Challoner PB, Cherry JM, Howard EA, Katzen AL, Pan W-C, Ryan T (1983): DNA termini in ciliate macronuclei. Cold Spring Harbor Symp Quant Biol 47:1195–1207.

Blackburn E, Challoner P, Cherry M, Howard E, Ryan T, Spangler E (1985): Genomic rearrangements in macronuclear development of Tetrahymena. In Herskowitz I, Simon M (eds): "Genome Rearrangement" (UCLA Symposia on Molecular and Cellular Biology, Vol 20, New Series). New York: Alan R. Liss, Inc., pp. 191–203.

Bollum FJ (1974) Terminal deoxynucleotidyl transferase. In Boyer PD (ed): "The Enzymes, Vol X, 3rd ed." New York: Academic Press, pp 145–171.

Dawson D, Herrick G (1982): Micronuclear DNA sequences of Oxtricha fallax homologous to the macronuclear inverted terminal repeat. Nucleic Acids Res 10:2911–2924.

Dugaiczyk A, Robberson DL, Ullrich A (1980): Single-stranded poly-(deoxyguanylic acid) associates into double- and triple-stranded structures. Biochemistry 19:5869–5873.

Emery HS, Weiner AM (1981): An irregular satellite sequence is found at the termini of the linear extrachromosomal rDNA in Dictyostelium discoideum. Cell 26:411–419.

Herrick G, Wesley RD (1978): Isolation and characterization of a highly repetitious inverted terminal repeat sequence from Oxytricha macronuclear DNA. Proc Natl Acad Sci USA 75:2626–2630.

Johnson EM (1980): A family of inverted repeat sequences and specific single-strand gaps at the termini of the *Physarum* rDNA palindrome. Cell 22:875–886.

Katzen AL, Cann GM, Blackburn EH (1981): Sequence-specific fragmentation of macronuclear DNA in a holotrichous ciliate. Cell 24:313–320.

King BO, Yao M-C (1982): Tandemly repeated hexannucleotide at *Tetrahymena* rDNA free end is generated from a single copy during development. Cell 31:177–182.

Klobutcher LA, Swanton MA, Donini P, Prescott DM (1981): All gene-sized DNA molecules in four species of hypotrichs have the same terminal sequence and an unusual 3′ terminus. Proc Natl Acad Sci USA 78:3015–3019.

Lillehaug JR, Kleppe RK, Kleppe K (1976): Phosphorylation of double-stranded DNAs by T$_4$ polynucleotide kinase. Biochemistry 15:1858–1865.

Oka Y, Shiota S, Nakai S, Nishida Y, Okubo S (1980): Inverted terminal repeat sequence in the macronuclear DNA of *Stylonychia pustulata*. Gene 10:301–306.

Pan W-C, Blackburn EH (1981): Single extrachromosomal ribosomal RNA gene copies are synthesized during amplification of the rDNA in *Tetrahymena*. Cell 23:459–466.

Pays E, Laurent M, Delinke K, Van Meirvenne N, Steinert M (1983): Differential size variations between transcriptionally active and inactive telomeres of *Trypanosoma brucei*. Nucleic Acids Res 11:8137–8147.

Pluta AF, Dani GM, Spear BB, Zakian VA (1984): Elaboration of telomeres in yeast: recognition and modification of termini from *Oxtricha* macronuclear DNA. Proc Natl Acad Sci USA 81:1475–1479.

Prescott DM, Murti KG (1974) Chromosome structure in ciliated protozoans. Cold Spring Harbor Symp Quant Biol 38:609–618.

Roychoudhury R, Jay E, Wu R (1976): Terminal labeling and addition of homopolymer tracts to duplex DNA fragments by terminal deoxynucleotidyl transferase. Nucleic Acids Res 3:101–116.

Saenger W (1983): "Principles of Nucleic Acid Structure." New York: Springer-Verlag, Chapter 13.

Shampay J, Szostak JW, Blackburn EH (1984): DNA sequences of telomeres maintained in yeast. Nature 310:154–157.

Szostak JW, Blackburn EH (1982): Cloning yeast telomeres on linear plasmid vectors. Cell 29:245–255.

Tu C-PD, Cohen SN (1980): 3′ end labeling of DNA with [α-^{32}P]cordycepin-5′-triphosphate. Gene 10:177–183.

Van der Ploeg LHT, Liu AYC, Borst P (1984): Structure of the growing telomeres of trypanosomes. Cell 36:459–468.

Williams RO, Young JR, Majiwa PAO (1982): Genomic location of T. brucei VSG genes: Presence of a minichromosome. Nature 299:417–421.

Yao MC, Yao CH (1981): The repeated hexanucleotide C-C-C-C-A-A is present near free ends of macronuclear DNA of *Tetrahymena*. Proc Natl Acad Sci USA 78:7436–7439.

III. Transformation and Gene Expression in *Drosophila*

Molecular Developmental Biology, pages 85–101
© 1986 Alan R. Liss, Inc.

Studies on the Developmentally Regulated Amplification and Expression of *Drosophila* Chorion Genes

Fotis C. Kafatos, Christos Delidakis, William Orr,
George Thireos, Katia Komitopoulou, and Yuk-Chor Wong

Department of Cellular and Developmental Biology, Harvard University,
Cambridge, Massachusetts 02138 (F.C.K., C.D., W.O., Y.-C.W.), Institute of
Molecular Biology and Biotechnology and Department of Biology, University of
Crete, 711 10 Heraclio, Crete, Greece (F.C.K., G.T.), and Department of
Biochemistry, Cell and Molecular Biology and Genetics, University of Athens,
Panepistimiopolis, Kouponia, Athens 15701, Greece (K.K.)

I. INTRODUCTION

The insect eggshell or chorion has been studied extensively as a model system for the study of programmed, differential gene expression during development [reviewed in Kafatos, 1983; Regier and Kafatos, 1985; Goldsmith and Kafatos, 1984]. At the end of oogenesis, the chorion proteins are synthesized by the follicular epithelial cells and secreted onto the surface of the oocyte, where they assemble to form the eggshell. Multiple strings of follicles (ovarioles) are present in the ovaries; because each constitutes a developmental series of progressively more mature follicles, tissues representing the various stages of choriogenesis can be isolated by simple dissection. In follicles of successive choriogenic stages, synthesis of specific chorion proteins "flows and ebbs" as the corresponding mRNAs accumulate and then disappear. Thus expression of the chorion structural genes occurs in the follicular epithelial cells according to a developmental program, which is amenable to analysis by molecular methods and, in *Drosophila melanogaster*, by genetics and germline transformation as well.

The ovary of *D. melanogaster* consists of many parallel ovarioles, each containing a single file of follicles (typically six or seven at different stages of development). Choriogenesis occurs in the final stages of follicle development (stages 11–14) [King, 1970), when two major structural components, endochorion and exochorion, are successively laid down by the follicle cells. These chorion layers (probably mostly the endochorion) are composed of a set of six major and more than 14 minor proteins, which can be resolved by two dimensional gel electrophoresis; subsets of these proteins are expressed in a temporally regulated mode during the 5 hr of choriogenesis [Petri et al., 1976; Waring and Mahowald, 1979; Margaritis et al., 1980].

Thus far, molecular analysis has defined two clusters of genes encoding major chorion proteins: one on the X chromosome at 7F1-2 and one on the third chromosome at 66D12-15 [Spradling et al., 1980; Griffin-Shea et al., 1980, 1982; Spradling, 1981]. The gene cluster at 7F1-2 codes for proteins s36-1 and s38-1, as well as several additional unidentified chorion components, whereas the one at 66D12-15 encodes the four lower-molecular-weight proteins s15-1, s16-1, s18-1, and s19-1. Each gene is represented by a single copy in the haploid genome.

Follicle cells undergo endomitotic replications between stages 7 and 11 of oogenesis, reaching a ploidy level of about 45C [Mahowald et al., 1979]. Overlapping this endomitosis is a differential replication (amplification) of both major chorion gene clusters, which is necessary for the extensive synthesis of the corresponding proteins [Spradling and Mahowald, 1980]. Amplification starts in stage 8–9 follicles, thus preceding choriogenesis, and reaches maximum levels of 20-fold on the X and more than 80-fold on the third chromosome. This process results on both chromosomes in a multiforked ("onionskin") replication structure [Spradling, 1981; Osheim and Miller, 1983]. The importance of amplification is indicated by the observation that the underamplification consequent to the *oc* inversion, which disrupts the X-chromosome cluster, is accompanied by underproduction of specific chorion proteins and morphological abnormalities of the eggshell [Spradling et al., 1979; Spradling, 1981]. Similarly, as will be discussed below, *trans*-acting mutants, which suppress amplification, correspondingly reduce chorion mRNA and protein production and consequently disrupt chorion formation [Orr et al., 1984].

Although the genes clustered within a locus are amplified equally and coordinately, careful analysis of the third chromosome cluster has established that the expression of its four genes is subject to quantitative and temporal controls that are gene-specific [Griffin-Shea et al., 1982]. Despite their equal gene dosage, the genes are expressed into mRNA in amounts that differ severalfold. Furthermore, although all four genes are predominantly expressed late in choriogenesis (stages 13 and 14), they differ somewhat in their developmental kinetics of mRNA accumulation and disappearance (and, correspondingly, in the kinetics of synthesis of the respective proteins). It is assumed that these temporal specificities are based on transcriptional differences, although differences at the level of mRNA processing and stability have not been excluded. Regulation is also gene-specific in the X-chromosome chorion gene cluster [Parks and Spradling, 1981].

In summary, choriogenesis in *D. melanogaster* entails phenomena that are both tissue specific and time specific: differential amplification of the chorion genes and their regulated expression into mRNA. Analysis of both *cis* and *trans* regulation is feasible and desirable. *Cis* regulation can be studied by reverse genetic procedures that are becoming standard: cloning and sequence analysis of the genes, in vitro engineering of the cloned DNA, and functional analysis of the modified constructs after germline transformation. *Trans* regulation can be studied through direct genetics, i.e., isolation and characterization of chorion-defective mutants that map away from the structural genes. The genetic studies are of special biological interest; they are likely to identify previously unsuspected functions and ultimately illuminate the logic of the developmental program as a whole.

II. SEQUENCE ANALYSIS OF CHORION GENES

We have sequenced a 4.8 kb genomic DNA fragment containing the three chorion genes known as s15-1, s18-1, and s19-1 [Wong et al., 1985]. For present purposes, the important conclusions are as follows: 1) The three chorion genes have a simple and consistent structure, composed of one small and one large exon separated by a small intron. The genes are transcribed from the same strand, and are 790–870 nucleotides apart. 2) The coding regions show limited similarities, suggesting that the chorion genes constitute an extensively diverged gene family. Divergence is so extensive that the genes do not cross-hybridize by normal criteria, and their flanking sequences can be compared without interference from matches of purely historical significance. 3) An intriguing feature of the flanking sequences is the occurrence of long T-rich regions, usually preceded by A-rich regions at a constant position relative to the genes: centered at 410–460 nucleotides upstream of each cap site or 350–390 nucleotides downstream of the 3' end. These regions show a striking prevalence of short, imperfect symmetries, such as palindromes or inverted repeats whose complement is found within 200 bp. This feature suggests the possible prevalence of protein binding sites. The density of symmetries is particularly high in the AT-rich region between s18-1 and s15-1. 4) Only short direct repeats are found in the 4.8 kb of DNA, but some of them are statistically significant and might be functionally important. The most intriguing case is the recurrence of short sequences near the 5′ ends of the genes in the region from −46 to −107 (Fig. 1). That region encompasses the hexanucleotide TCACGT, which is found at equivalent positions in all three genes. Another interesting element of the same region is GAGCCGAAAC, which is matched at least in part by each of the genes and is adjacent to an inverted repeat that differs in sequence but not location. It is not unlikely that such sequence features will prove important for the regulated expression of the chorion genes.

III. *CIS* REGULATION OF AMPLIFICATION AND GENE EXPRESSION

Amplification of the chorion gene clusters extends into 90–100 kb chromosomal domains [Spradling, 1981]. By using a series of probes specific for outlying regions, Spradling was able to demonstrate that the differential replication extends to 40–50 kb of DNA to either side of each chorion gene cluster and is not uniform but exhibits gradients of dosage decreasing with distance from the centrally located chorion genes. These and other data are consistent with the following model. Differential replication starts in the gene cluster itself (possibly at a single origin) and proceeds bidirectionally to a variable distance, forming a bubble with two replication forks at its ends. The origin(s) reinitiates, and continuation of this cycle leads to a multiforked

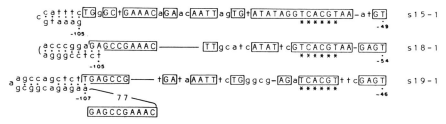

Fig. 1.   Similarities in the 5′ flanking regions of three sequenced chorion genes. Numbers refer to distance from the putative cap site. Note the invariably conserved TCACGT element (asterisks). Matched bases are capitalized and boxed, mismatches are in lower case, and deletions are represented by dashes or horizontal lines. At the left of each sequence, an inverted repeat is shown as a hairpin structure, with dots indicating complementary bases. A perfect match to the GAGCCGAAAC element typified by s18-1 is also found 77 nucleotides upstream of the inverted repeat in s19-1. From Wong et al. [1985], with permission.

replication structure. Indeed, such structures have been visualized in chromatin spreads of stage 11 and 12 follicle cells [Osheim and Miller, 1983].

The effects of the *oc* inversion demonstrated that *cis*-acting DNA elements are necessary for amplification and can lead to amplification of other sequences brought into their vicinity [Spradling et al., 1979; Spradling and Mahowald, 1981]. As a first step in the analysis of such elements, we used the yeast ARS assay [Stinchcomb et al., 1979] to search for DNA replication origins in 12 kb of DNA encompassing the chorion genes of the third chromosome cluster and in 18 kb from the X chromosome cluster (Thireos et al., in preparation). *Drosophila* chromosomal DNA fragments were cloned in either the EcoRI or the HindIII sites of YIp5, a pBR322 derivative carrying the yeast *ura3* gene, and were used to transform the ura3-52 strain of yeast, which carries a stable mutation of *ura3*. Any *Drosophila* ARS sequence, i.e., a sequence capable of autonomous replication in yeast, would allow transformed strains to grow on minimal media. Only one region with ARS activity was detected in each chorion cluster. Figure 2 shows the one from the third chromosome, which is localized to a 0.75 kb EcoRI-SacI fragment upstream of the s16-1 gene. In *Drosophila* embryos, replicons average 8 kb in length, whereas in tissue culture cells and more slowly growing larval tissues they range up to 15-fold longer [Blumenthal et al., 1973; Zakian, 1976; Ananiev et al., 1977; Steinemann, 1981a,b]. Thus these ARS sequences could be actual replication origins, although they could also be insignificant; the assay system is heterologous.

With the development of the P element-mediated transformation procedure [Spradling and Rubin, 1982; Rubin and Spradling, 1982], both the Spradling lab and ours undertook more direct tests for amplification-controlling elements. Results from the Spradling lab have been published recently [deCicco and Spradling, 1984] and are summarized elsewhere in this volume. Our results are summarized in Figure 2, and typical data are presented in Figure 3.

Overlapping fragments of the third chromosome cluster were inserted into the genome of the recipient strain $cn;ry^{42}$ by P element transformation and tested for their ability to amplify. The inserted chorion sequence was immediately juxtaposed to the 7.2 kb HindIII fragment carrying a wild type ry^+ gene, which was used as a marker to screen for transformants. Amplification of the inserted sequences was assayed by isolating DNA from ovaries or late-stage follicles of females from the transformed lines and measuring in Southern blots the relative intensities of bands that correspond to the insert vs. bands that correspond to endogenous, unamplified, unrelated sequences and to fully amplified, endogenous chorion genes. DNAs from males were

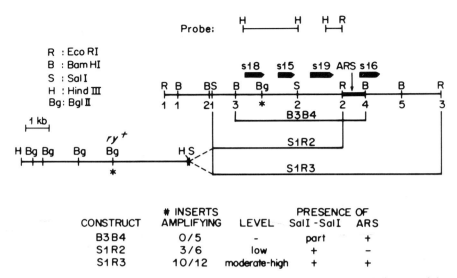

Fig. 2. Restriction map of the third chromosome chorion gene cluster and diagram of the constructs used for studies on amplification. The positions of the four major chorion transcripts and the ARS element are also shown. The indicated subclones (B3B4, S1R2, S1R3) were inserted adjacent to *ry*+ DNA in the P element Carnegie 20 vector [deCicco and Spradling, 1984] and used for transformation. The chorion-specific probe used in Figure 3 is indicated, and asterisks outline the T1 fragment seen in Figure 3. The table at bottom summarizes the results obtained with a total of 23 independent inserts. The second column shows the number of transformed lines that amplify out of the total number of lines obtained with that construct. The third column indicates the level of amplification (see Text).

used as unamplified controls. Whenever amplification of the insert occurred, it was always tissue-specific, i.e., occurred only in the ovaries and never in ovariectomized females. Furthermore, the *ry*+ sequence adjacent to the transformed chorion fragment also showed amplification in all amplifying inserts.

For convenience, the chorion DNA fragments will be designated by reference to rare-cutter restriction endonuclease sites that define their borders. We have tested for amplification (Fig. 2) fragments B3B4 (5.65 kb), S1R2 (5.75 kb), and S1R3 (10.1 kb). In agreement with the results of deCicco and Spradling [1984], the B3B4 construct invariably failed to amplify. That construct includes the ARS sequence but is missing part of the 3.8 kb SalI fragment (S1S2 in Fig. 2), which deCicco and Spradling [1984] have defined as a minimal amplification-control element (ACE). By contrast, amplification is sometimes observed in the S1R2 construct, which is similar in size to B3B4 but includes the entire S1S2 ACE element although lacking the ARS

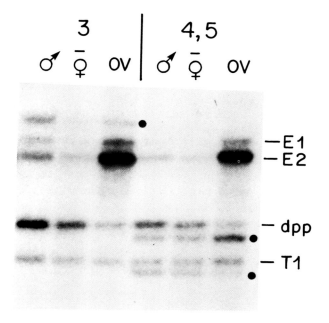

Fig. 3.   A typical amplification blot. DNAs were prepared from two transformed lines, S1R2-3 and S1R2-4/5, the latter bearing two independent inserts. The DNAs were from males (♂), ovariectomized females (♀), and ovaries (OV). After digestion with BgIII, Southern hybridization was performed with a mixture of probes from the third chromosome chorion cluster (see Fig. 2) and from a single-copy, nonamplified control region (part of the *dpp* complex at 22F1-2; courtesy D. St. Johnston). The *dpp* band permits normalization for the amount of DNA loaded, which was substantially lower in the ovary. E1 and E2 are the endogenous chorion bands and clearly show ovary-specific amplification. T1 is a transformant-specific chorion plus ry^+ band internal to the insert (cf. asterisks in Fig. 2). Transformant-specific junction fragments that span chorion DNA and DNA from the integration site are indicated by dots. When the levels of amplification of transformant-specific and endogenous gene-specific fragments were measured in late follicles (stages 12–14), they were estimated as 6% for S1R2-3 and as 3% and 11% for the two inserts of S1R2-4,5, thus indicating substantial chromosomal position effects on amplification.

sequence. These results indicate that the ARS element is inessential and the ACE element is essential for amplification. However, it is notable that even the S1R2 construct gives only a weak amplification response: Only three of six independent transformants amplify, and the extent of amplification is low (three- to 15-fold), relative to the endogenous chorion gene cluster (80-fold).

The results with S1R3 are in marked contrast. In this case, successful amplification is more frequent (10/12 independent transformants), and, more importantly, the typical extent of amplification is moderate or high, ranging up to 80-fold; approximately half of the transformants amplify to more than half the level of the endogenous genes ($\geqslant$40-fold). The only difference between the S1R2 and S1R3 constructs is the presence in the latter of a 4.35 kb EcoRI fragment that contains the ARS sequence. The results must be interpreted cautiously, however, because in the experiments of deCicco and Spradling [1984] the presence of additional DNA sequences attached to the ACE fragment modified the latter's amplification response.

Another striking and possibly related observation is that amplification is extremely sensitive to chromosomal position: The response of the same fragment varies widely at different insertion sites. The position effect is apparent in Figure 3, in which two independent inserts of the same construct in the same ovaries are seen to differ extensively (fivefold) in the level of amplification. The results of deCicco and Spradling [1984] also indicate extensive position effects.

It is too early to interpret these observations definitively. It is possible that the only significant *cis*-acting elements are found in the ACE fragment and that the beneficial effect of additional DNA sequences in the constructs is due merely to "buffering" the ACE functions against some type of site-specific interference. Alternatively, it is possible to speculate that multiple *cis*-acting elements are needed for amplification, that the ACE region contains one or more such elements that cannot be supplied by other chromosomal sites, and that an ARS element is also needed but can be provided by the insertion site. For example, the replication origin used in amplification might be nonspecific and found outside the ACE region (ARS sequence?), whereas ACE might contain one or more essential elements that recognize a specific replication-activating factor unique to choriogenic cells. Further experiments are in progress to discriminate between these possibilities.

Entirely analogous experiments are underway for characterizing the *cis*-acting DNA elements that regulate the tissue and temporal specificity of chorion gene expression. The approach is to define the minimal DNA sequence that permits normal expression of a chorion gene integrated into the germline via a P element vector and then to alter that sequence systematically by deletions and nucleotide substitutions to define the actual DNA elements involved. We have focused on the s15-1 gene, which shows a particularly interesting temporal specificity in vivo [Griffin-Shea et al., 1982]. Transformants have been produced with contructs that bear variable portions of the s15-1 5′ flanking sequence connected to two types of reporter genes: the alcohol dehydrogenase gene or s15-1 itself modified by the in-frame insertion of part of a noncross-hybridizing silkmoth chorion gene

(Mariani, Lingappa and Romano, personal communication). Assays of the patterns of expression of these transformed genes are currently in progress.

IV. *TRANS* REGULATION OF CHORIOGENESIS

We have begun a genetic effort to dissect the various steps in the program of choriogenesis. Many chorion mutants are expected to lead to infertility of the eggs, i.e., to female sterility, a trait for which systematic mutant searches have been conducted [Bakken, 1973; Gans et al., 1975; Mohler, 1977]. Through a new series of ethylmethane sulfonate (EMS) mutagenesis experiments, conducted in collaboration with M. Gans at Gif-sur-Yvette, we expanded the collection of female-sterile mutants that map on the X chromosome and studied in more detail eggshell mutants [Komitopoulou et al., 1983; Komitopoulou and Kouyanou, in preparation]. A total of 69 recessive mutants were assigned to 28–31 complementation groups, of which 21 were partially characterized by ultrastructural and biochemical procedures. Of the latter, nine showed prominent ultrastructural defects in the chorion, at least in eight cases accompanied by deficiencies in characterized chorion proteins. Interestingly, some of the mutants were partially fertile. The cytogenetic locations of chorion-defective mutants are shown in Figure 4 . At least six complementation groups map far from the known chorion structural genes.

Two of these mutants, K*451* and K*1214,* were subjected to detailed molecular analysis [Orr et al., 1984]. Their prominent ultrastructural defects are accompanied by substantial underproduction of all six major chorion protein bands, s15, s16, s18, s19, s36, and s38 (Komitopoulou et al., in preparation). This underproduction of the proteins was traced to a corresponding underaccumulation of the pertinent mRNAs; the temporal specificities of the mRNAs were not affected, however (Fig. 5). The latter observation clearly indicates that the mutants do not have a trivial explanation, such as cell death. In turn, the reduced mRNA levels were attributed to a correspondingly reduced gene dosage resulting from a substantial suppression of amplification in both the X and the third chromosome (Fig. 6). Because the genes in question map far from or even in a different chromosome than the chorion gene clusters, the suppression of amplification operates in *trans.* More recently, four additional chorion-defective mutants, two from the X and two from the third chromosome, were shown to reduce amplification in *trans* (Komitopoulou, Kouyanou, Snyder and Galanopoulos, unpublished observations). In that the majority of the six characterized complementation groups that affect amplification are known from single alleles, and insofar as additional unlinked mutants are candidates for similar defects (unpublished obser-

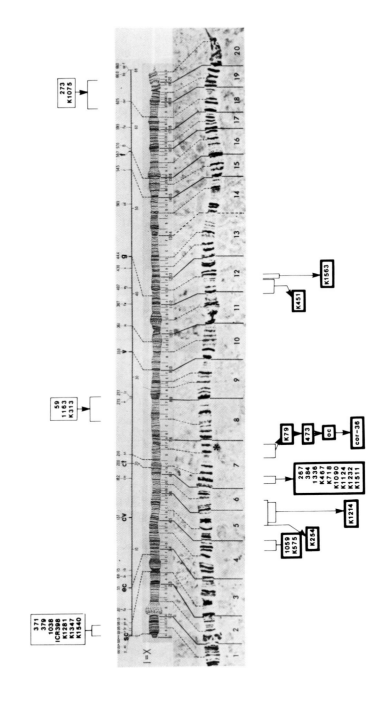

vations), it is clear that many genetic functions are necessary for normal amplification. This observation might be related to the remarkable degree to which amplification is subject to position effects (see above).

In general, the functions necessary for normal amplification might be quite diverse, covering a very wide spectrum from nonspecific to specific or indirect to direct. At one extreme might be mutants in the general replication machinery, which is known to be complex even in procaryotes [Kornberg, 1982; Alberts et al., 1983]: It might be that amplification places especially heavy demands on the replication process so that leaky and therefore viable general replication mutants (or mutants in redundant replication functions) are scored as amplification-defective. At the other extreme would be mutants of highly specific diffusible factors that act directly on the amplification-control elements, e.g., initiation-promoting proteins that bind to the amplification origins and single them out for differential replication. In between these extremes would be mutants in functions that are indirect but developmentally interesting; for example, functions involved in intercellular coordination between oocyte and follicle cells, intracellular functions that relate the program of amplification to other aspects of follicular cell development (such as the sequential expression of genes for yolk proteins, vitelline membrane, and chorion), and functions that might affect amplification through their influence on local chromosomal conditions, such as the state of chromatin condensation. Each type of mutant would be of some intrinsic interest, but naturally we are most interested in those with developmental significance. At present, we know only that K451 is allelic to *mus-101* (Baker, personal communication), a mutant originally selected for sensitivity to mutagens [Boyd et al., 1976]. The *mus-101* allele apparently shows some defects in postreplication repair and leads to chromosomal loss and nondisjunction [Boyd et al., 1976; Boyd and Setlow, 1976; Baker and Smith, 1979]. Interestingly, *mus-101* also leads to failure of mitotic condensation of heterochromatin, although not euchromatin, and the wild type gene action extends into phase G2 (i.e., is not limited to the time of DNA synthesis [Gatti et al.,

---

Fig. 4.   Localization of female-sterile mutants. Complementation groups that show minor or prominent chorion abnormalities are shown with thin and thick borders, above and below the chromosomal map, respectively. For the mutants (K79,473, oc, and *cor-36*) that map in the vicinity of the known s36-1 and s38-1 chorion genes (asterisk), complementation analysis has not been performed. K499 also shows prominent chorion abnormalities but has not been mapped. Note that five genes (K575, K254, K1214, K451, K1563) that affect both chorion structure and the abundance of major chorion proteins map at a distance from the corresponding structural genes. From Komitopoulou et al. [1983], with permission.

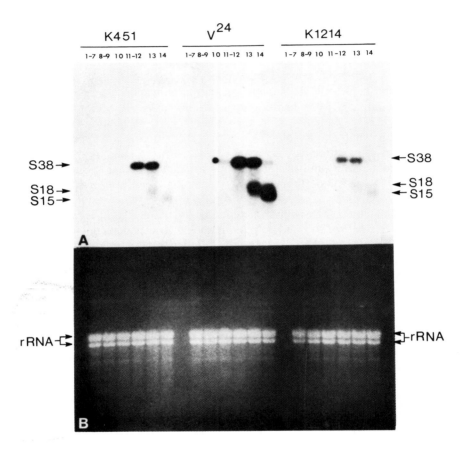

Fig. 5.   Accumulation of chorion mRNAs during development in control (v^{24}) and homozygous mutant strains. Equal amounts of total nucleic acids were isolated from follicles of the indicated stages, and the RNAs were analyzed by RNA transfer blotting using a mixture of probes corresponding to genes s38-1, s18-1, and s15-1. Note the approximate equality of RNA in all samples (B) monitored by the intensity of the rRNA bands. Also note in A the normal developmental specificities but underaccumulation of the chorion mRNAs in the mutants and the more severe effect on mRNAs encoded by the third chromosome gene cluster (s18-1 and s15-1). From Orr et al. [1984], with permission.

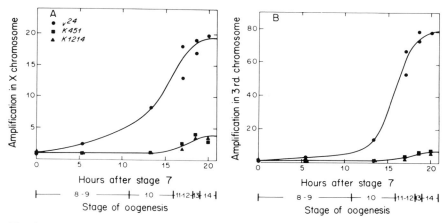

Fig. 6.   Densitometric estimates of the time course of amplification in the two chorion loci in the wild type ($v/^{24}$) and two homozygous mutants. Data are from Southern blot analyses in which staged follicle DNAs were blot hybridized with a mixture of probes representing the two chorion gene loci and two control, unamplified loci. The extent of amplification was evaluated relative to the internal unamplified controls. Times are calculated from published data [King, 1970]. A, time course of amplification in the X chromosome; B, time course of amplification in the third chromosome. The symbols used are as in A. From Orr et al. [1984], with permission.

TABLE I. Dysgenic X Chromosome Mutants

Class	Description	Egg phenotype	Fertility	Larval development	No. of lines
I	Sterile abnormal eggs	−	−	−	105
II	Embryonic lethals	+	−	−	31
III	Fertile abnormal eggs	−	+	+	82
IV	No eggs laid	NA	NA	NA	114
V	Larval lethals	+	+	−	10
VI	Low fecundity	+	+	+	3
VII	Homozygous lethals/ low viability	NA	NA	NA	25
VIII	Homo- and hemizygous lethals/ low viability	NA	NA	NA	5

1983]). Further characterization is necessary for identification of the developmentally most interesting amplification mutants.

Additional choriogenesis-defective mutants that have not been identified as yet should include those that affect in *trans* the regulated transcription of chorion genes as well as those that correspond to minor chorion structural genes. Irrespective of their nature, analysis of the mode of action of choriogenesis mutants would be considerably facilitated by our ability to clone the respective genes. A useful approach to the molecular cloning of genetic loci for which products are not known is to generate "tagged" alleles via P-M hybrid dysgenesis [Kidwell et al., 1977; Bingham et al., 1982; Rubin et al., 1982; Searles et al., 1982]. This entails mating P-strain males to M-strain females, leading to "dysgenic hybrid" progeny. Flies of the P strain are characterized by the presence of P elements dispersed throughout their chromosomes; M-strain flies lack intact P elements. P elements are normally stable in a P-strain cytoplasmic background (P cytotype). However, if P element-containing chromosomes are exposed to an M-strain ooplasm, the P elements undergo transposition. This destabilization occurs principally in the dysgenic hybrid germline and results in the insertion of P elements into new chromosomal sites. Such sites can then readily be cloned by hybridizing a P element probe to the appropriate recombinant libraries.

We have carried out a hybrid dysgenic screen of the X chromosome in an effort to obtain tagged mutations that identify structural and regulatory elements involved in eggshell formation. Over 4,000 dysgenic lines were examined for both fertility and egg morphology, and 375 lines were selected (Table I), representing several mutant classes, which occasionally overlap. The class I females are sterile or only slightly fertile and lay eggs that have an aberrant morphology. Preliminary ultrastructural analysis indicates that some of the aberrant eggs have prominent chorion defects (Galanopoulos and Snyder, unpublished observations). It remains to be determined by biochemical analysis whether these defects reflect abnormal or deficient chorion gene expression. Like mutants of the first category, the class III mutants lay eggs that are morphologically abnormal. However, these mutants exhibit nearly normal levels of fertility. Previous studies have shown that some mutants of this type can correspoond to chorion defects [Komitopoulou et al., 1983]. The class IV mutants are completely sterile, laying no eggs. The ovaries of these flies have been examined to ascertain at what state during oogenesis defects appear. In 22 of the mutants that have been dissected to date, oocyte development proceeds normally up to discrete stages of choriogenesis (stages 11–14), after which their development is arrested or abnormal. We are testing these dysgenic mutants for allelism to the K-series mutants with known chorion defects, including the amplification-deficient mutants K*451* and K*1214*, and are characterizing them morphologically and biochemically for independent chorion defects. We expect to generate an expanded repertoire of mutations that possess their own molecular tags and that act in *cis* or in *trans*, directly or indirectly, on chorion gene amplification and expression.

ACKNOWLEDGMENTS

We thank our former and present associates, especially V. Galanopoulos, A. Georgi, S. Kouyanou, J. Lingappa, B. Mariani, J. Pustell, P. Snyder, and N. Spoerel, for their contributions to the work reviewed here-in and for permission to quote unpublished results. The work was supported by grants from the NIH, the American Cancer Society, and the Greek Ministry for Research and Technology.

V. REFERENCES

Alberts BM, Barry J, Bedinger P, Formosa T, Jongeneel CV, Kreuzer KN (1983): In vitro T4 DNA replication. Cold Spring Harbor Symp Quant Biol 47:655–668.

Ananiev EV, Polukarova LG, Yurov YB (1977): Replication of chromosomal DNA in Drosophila melanogaster cells cultured in vitro. Chromosoma 59:259–272.

Baker BS, Smith DA (1979): The effects of mutagen-sensitive mutants of Drosophila melanogaster in nonmutagenized cells. Genetics 92:833–847.

Bakken AH (1973): A cytological and genetic study of oogenesis in Drosophila melanogaster. Dev Biol 33:100–122.

Bingham PM, Kidwell MG, Rubin GM (1982): The molecular basis of P-M hybrid dysgenesis: The role of P element, a P-strain-specific transposon family. Cell 29:995–1004.

Blumenthal AB, Kriegstein HJ, Hogness DS (1973): The units of DNA replication in Drosophila melanogaster chromosomes. Cold Spring Harbor Symp quant Biol 38:205–223.

Boyd JB, Golino M, Nguyen T, Green MM (1976): Isolation and characterization of X-lined mutants of Drosophila melanogaster which are sensitive to mutagens. Genetics 84:485–506.

Boyd JB, Setlow RB (1976): Characterization of post replication repair in mutagen-sensitive strains of Drosophila melanogaster. Genetics 84:507–526.

deCicco DV, Spradling AC (1984): Localization of cis-acting element responsible for the developmentally regulated amplification of Drosophila chorion genes. Cell 38:45–54.

Gans M, Audit C, Masson M (1975): The isolation and characterization of sex-linked female-sterile mutants in Drosophila melanogaster. Genetics 81:683–704.

Gatti M, Smith DA, Baker BS (1983): A gene controlling condensation of heterochromatin in Drosophila melanogaster. Science 221:83–85.

Goldsmith MR, Kafatos FC (1984): Developmentally regulated genes in silkmoths. Annu Rev Genet 18:443–487.

Griffin-Shea R, Thireos GC, Kafatos FC (1982): Organization of a cluster of four chorion genes in Drosophila and its relationship to developmental expression and amplification. Dev Biol 91:325–336.

Griffin-Shea R, Thireos G, Kafatos FC, Petri WH, Villa-Komaroff L (1980): Chorion cDNA clones of Drosophila melanogaster and their use in studies of sequence homology and chromosomal location of chorion genes. Cell 19:915–922.

Kafatos FC (1983): Structure, evolution and developmental expression of the chorion multi-gene families in silkmoths and *Drosphila*. In Subtelny S, Kafatos FC (eds): "Gene Structure and Regulation in Development." New York: Alan R. Liss, Inc., pp 33–61.

Kidwell MG, Kidwell JF, Sved JA (1977): Hybrid dysgenesis in Drosophila melanogaster: A syndrome of aberrant traits including mutation, sterility and male recombination. Genetics 86:813–833.

King RC (1970): "Ovarian Development in Drosophila melanogaster." New York: Academic Press.

Komitopoulou K, Gans M, Margaritis LH, Kafatos FC, Masson M (1983): Isolation and characterization of sex-linked female-sterile mutants in Drosophila melanogaster with special attention to eggshell mutants. Genetics 105:897–920.

Kornberg A (1982): "Supplement to DNA Replication." San Francisco: W.H. Freeman.

Mahowald AP, Coulton JH, Edwards MK, Floyd AF (1979): Loss of centrioles and polyploidization in follicle cells of Drosophila melanogaster. Exp Cell Res 118:404–410.

Margaritis LH, Kafatos FC, Petri WH (1980): The eggshell of Drosophila melanogaster. I. Fine structure of the layers and regions of the wild-type eggshell. J Cell Sci 43:1–35.

Mohler JD (1977): Developmental genetics of the Drosophila egg. I. Identification of 59 sex-linked cistrons with maternal effects on embryonic development. Genetics 85:259–272.

Orr W, Komitopoulou K, Kafatos FC (1984): Mutants suppressing in *trans* chorion gene amplification in Drosophila. Proc Natl Acad Sci USA 81:3773–3777.

Osheim YN, Miller OL Jr (1983): Novel amplification and transcriptional activity of chorion genes in Drosophila melanogaster follicle cells. Cell 33:543–553.

Parks S, Spradling AC (1981): The temporal program of chorion gene expression. Carnegie Inst Wash Year Book 80:188–191.

Petri WH, Wyman AR, Kafatos FC (1976): Specific protein synthesis in cellular differentiation. III. The eggshell proteins of *Drosophila melanogaster* and their program of synthesis. Dev Biol 49:185–199.

Regier JC, Kafatos FC (1985): Molecular aspects of chorion formation. In Kerkut GA, Gilbert LI (eds): "Embryonogenesis and Reproduction: Comprehensive Insect Physiology, Biochemistry and Parmacology, Vol 1." Oxford: Pergamon Press, pp 113–151.

Rubin GM, Kidwell MG, Bingham PM (1982): The molecular basis of P-M hybrid dysgenesis: The nature of induced mutations. Cell 19:987–994.

Rubin GM, Spradling AC (1982): Genetic transformation of Drosophila with transposable element vectors. Science 218:348–353.

Searles LL, Jokerst RS, Bingham PM, Voelker RA, Greenleaf AL (1982): Molecular cloning of sequences from a Drosophila RNA polymerase II locus by P element transposon tagging. Cell 31:585–592.

Spradling AC (1981): The organization and amplification of two chromosomal domains containing Drosophila chorion genes. Cell 27:193–201.

Spradling AC, Digan ME, Mahowald AP, Scott M, Craig EA (1980): Two clusters of genes for major chorion proteins of Drosophila melanogaster. Cell 19:905–914.

Spradling AC, Mahowald AP (1980): Amplification of genes for chorion proteins during oogenesis in Drosophila melanogaster. Proc Natl Acad Sci USA 77:1096–1100.

Spradling AC, Mahowald AP (1981): A chromosome inversion alters the pattern of specific DNA replication of Drosophila follicle cells. Cell 27:203–209.

Spradling AC, Rubin GM (1982): Transposition of cloned P elements into Drosophila germ line chromosomes. Science 218:341–347.

Spradling AC, Waring GL, Mahowald AP (1979): Drosophila bearing the *ocelliless* mutation underproduce two major chorion proteins both of which map near this gene. Cell 16:609–616.

Steinemann M, (1981a): Chromosomal replication in Drosophila virilis II. Organization of active origins in diploid brain cells. Chromosoma 82:267–288.

Steinemann M, (1981b): Chromosomal replication in Drosophila virilis III. Organization of active origins in the highly polytene salivary gland cells. Chromosoma 82:289–307.

Stinchcomb DT, Struhl K, Davis RW (1979): Isolation and characterization of a yeast chromosomal replicator. Nature 282:39–43.

Waring GL, Mahowald AP (1979): Identification and time of synthesis of chorion proteins in Drosophila melanogaster. Cell 16:599–607.

Wong Y-C, Pustell J, Spoerel M, Kafatos FC (1985): Coding and potential regulatory sequences of a cluster of chorion genes in *Drosophila melanogaster*. Chromosoma 92:124–135.

Zakian VA (1976): Electron microscopic analysis of DNA replication in main band and satellite DNAs of *Drosophila virilis*. J Mol Biol 108:305–331.

Molecular Developmental Biology, pages 103–114
© 1986 Alan R. Liss, Inc.

The *Drosophila melanogaster* Dopa Decarboxylase Gene: Progress in Understanding the In Vivo Regulation of a Higher Eukaryotic Gene

J. Hirsh

Department of Biological Chemistry, Harvard Medical School,
Boston, Massachusetts 02115

I. INTRODUCTION

In this brief review, we will discuss progress that has been made in understanding the developmental regulation of an enzyme-encoding gene in the fruit fly *Drosophila melanogaster*. This gene, *Ddc*, encodes the enzyme dopa decarboxylase, an enzyme with multiple functions in *Drosophila*. We will present a brief summary of the biology and genetics of this gene, its regulation during development, and preliminary data regarding the cis-acting sequences responsible for its regulation. Finally, we discuss the potential of using gene reintegration techniques to alter the physiology of eukaryotic organisms.

II. BACKGROUND

A. The *Ddc* Gene

The gene encoding the enzyme dopa decarboxylase is a well studied locus in *Drosophila melanogaster*. *Ddc* was initially localized [Hodgetts, 1975] by segmental aneuploidy, a genetic technique that allows the dosage of small regions of the genome to be varied. Only one small chromosomal region showed strict dosage correlation with levels of *Ddc* enzyme activity. In collaboration with Wright's group [1976b], small deficiencies within this region were isolated that decreased *Ddc* enzyme levels twofold in heterozygotes. These deficiencies were subsequently used to screen for *Ddc* point mutants [Wright et al., 1976a]. Recessive mutations were isolated that were lethal when heterozygous with one of the deficiencies and that reduced *Ddc* enzyme levels when heterozygous for a wild type chromosome. Further evidence that this locus encodes the *Ddc* enzyme comes from the observation of reduced *Ddc* protein levels in several *Ddc* mutants [Clark et al., 1978] and from the isolation of *Ddc* mutant lesions that lead to the production of a temperature-sensitive enzyme [Wright et al., 1982].

The DNA encoding *Ddc* was isolated using a multistep screening protocol taking advantage of both its known regulation and its cytologic location on salivary polytene chromosomes [Hirsh and Davidson, 1981]. RNA was isolated from organisms at two developmental stages representing periods of relatively high or low *Ddc* enzyme activity. Probes derived from these RNAs were used to screen genomic phage λ recombinants. Phage that selectively hybridized to the probe from the high activity stage were screened by in situ hybridization, hybridizing polytene chromosome squashes with pools of ten phage probes at a time. One pool was shown to hybridize to the band 3761,2, within the region to which *Ddc* had been localized. The phage responsible for this hybridization was isolated and was shown to contain a segment that could hybrid select RNA encoding the *Ddc* protein. Further studies have localized a *Ddc* genetic lesion to lie within this region [Gilbert et al., 1984]. The molecular structure of *Ddc* is shown in Figure 1.

Genetic and molecular analyses have shown that *Ddc* is located within a dense cluster of genes, many of which share phenotypic relations to *Ddc* [Gilbert et al., 1984]. At least nine complementation groups lie within at most 100 kb of DNA adjacent to *Ddc*, and six of these are within at most 23 kb of DNA, an exceptionally high gene density for *Drosophila*. The biological role of this clustering is not apparent; P element-mediated gene reintegration shows that *Ddc* functions normally when reintegrated at chromosomal locations distant from its neighboring genes [Scholnick et al., 1983], and genetic studies similarly show that the lethal genes adjacent to *Ddc* can function when *Ddc* has been deleted [Gilbert et al., 1984].

B. Functions of *Ddc* Metabolites

The catecholamine metabolites produced by the *Ddc* enzyme function in two biological processes in *D. melanogaster*. The majority of the enzyme

RNA Length

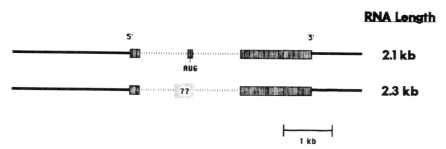

Fig. 1. Structure of *Ddc*. Exons are shown shaded in gray; introns are shown by hatching. The exons used to generate the 2.1 kb major form of the mRNA are shown in the top diagram. The bottom diagram shows the structure of exons leading to expression of a 2.3 kb RNA found exclusively during late embryogenesis [Beall and Hirsh, 1984]. The question marks show where the structure of this RNA differs from the 2.1 kb RNA. The sequence of nearly the entire gene has been determined (in preparation). The first AUG available for translational initiation is shown. The structures have been determined at low resolution by electron microscopic RNA:DNA hybrid analysis and by the use of specific probes [Beall and Hirsh, 1984] and at high resolution by primer extension mapping, S1 mapping, and by the isolation and sequencing of cDNA clones encoding the 2.1 kb RNA (Morgan and Hirsh, in preparation).

activity functions in a metabolic pathway leading to quinone compounds that participate in cross-linking and pigmentation of the cuticle [reviewed in Brunet, 1980]. The enzyme activity responsible for this function is located in the hypoderm directly beneath the developing cuticle. The effects of loss of this function are readily apparent in the phenotypic effects of *Ddc* mutant lesions. The lethality in homozygous *Ddc* mutant flies usually occurs at a molt, when cuticular synthesis is required. Furthermore, *Ddc* temperature-sensitive mutant lesions, which make only several percent of wild type levels of enzyme activity, show obvious structural and pigmentation defects of the cuticle. The pigmentation defects are most clear shortly after pupation: Mutant pupae are a yellow-green color, distinctly different from the red-brown coloration of wild type pupal cases. The clarity of this phenotypic difference allowed it to be used in detecting *Ddc*+ transformed flies in early P element-mediated transformation studies [Scholnick et al., 1983].

A second function for *Ddc* enzyme activity is found in the central nervous system, where the enzyme decarboxylates 5-hydroxytryptophan to form serotonin as well as decarboxylating L-dopa [Dewhurst et al., 1972; Livingstone and Tempel, 1983]. That the same gene is involved both in the hypodermal and the central nervous system *Ddc* activity is indicated by the finding that enzyme activity in the CNS is reduced by *Ddc* temperature-sensitive mutant lesions [Livingstone and Tempel, 1983]. Serotonin has been implicated as being involved in sensitization and classical conditioning of the siphon withdrawal reflex in *Aplysia* [reviewed briefly in Tempel et al., 1984]. Presumably serotonin and/or dopamine function as neurotransmitters in *Drosophila*, since *Ddc^{ts}* mutant flies are defective in learning [Tempel et al., 1984].

C. Regulation During Development

Ddc shows acute regulation during development. A developmental profile of *Ddc* enzyme activity is shown in Figure 2, showing expression of wild type genes and a gene reintegrated via P elements. Three major peaks of enzyme activity are apparent, occurring during late embryogenesis, just prior to pupation, and at adult eclosion. Each of these stages is a time when extensive cuticular synthesis, pigmentation, and hardening are taking place. Small increases in *Ddc* enzyme activity also occur at each of the larval molts [Kraminsky et al., 1980; Marsh and Wright, 1980].

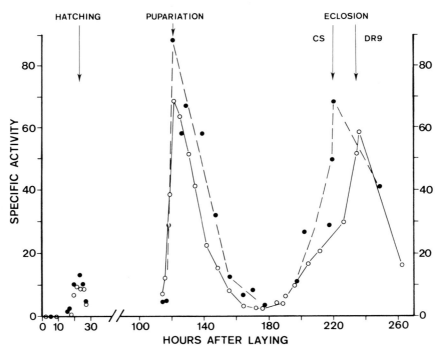

Fig. 2. Developmental profile of *Ddc* enzyme activity during development. Shown are profiles of a wild type Canton-S *Ddc* gene (solid circles) and a *Ddc* gene reintegrated into the chromosome via a P element vector (open circles). Reproduced from Scholnick et al. [1983], with permission.

As was mentioned previously, the same gene is responsible for expression of *Ddc* enzyme activity in the CNS. The data shown in Figure 3 demonstrate that the regulation of *Ddc* in the central nervous system is different from that in the hypoderm. CNS tissue was dissected from either early third instar larvae or from white prepupae, a stage at which *Ddc* enzyme levels are near their maxima. Enzyme levels in the CNS vary only slightly between these two stages, whereas activity in the remainder of the organism increases ca. tenfold. We have been unable to study the detailed structure of the *Ddc* transcript in the CNS owing to the low levels of transcripts and small amounts of this tissue.

It is not yet clear whether the central nervous system *Ddc* expression results from cell-specific expression of the gene or the expression occurs in the entirety of the tissue. White and Valles [1984] have used antibodies to serotonin to show that only a small subset of cells in the CNS shows immunoreactivity. This analysis cannot distinguish between cell specificity of the enzymes involved in serotonin synthesis from cell-specific transport of serotonin. Further studies utilizing antibodies to the *Ddc* protein will be necessary to distinguish between these alternatives.

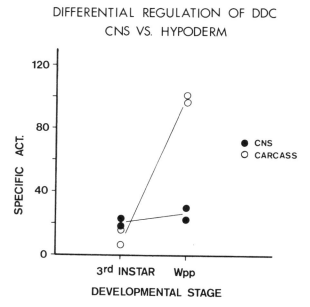

Fig. 3.   Differential regulation of *Ddc* in the central nervous system vs. the hypoderm. CNS tissue was hand dissected from wild type Canton-S organisms at early third instar or white prepupal stages. The activity in the carcass is largely owing to expression in the hypoderm. *Ddc* enzyme assays were performed as given in Scholnick et al. [1983].

The peak of *Ddc* enzyme activity just preceding pupation is well correlated with an increase in titers of the molting hormone β-ecdysone. A strain temperature-sensitive for the synthesis of β-ecdysone has been used to demonstrate that the enzyme activity at this stage is a relatively rapid response to exogenous addition of hormone [Kraminsky et al., 1980] or to returning to a temperature permissive for β-ecdysone synthesis [Marsh and Wright, 1980]. *Ddc* enzyme levels appear to respond within as little as 2 hr of exposure to hormone. Though the shortness of this time period has been used to argue that the gene is responding directly to the hormone [Kraminsky et al., 1980], no direct demonstration of such regulation has been published. If such regulation exists, it cannot explain the whole of *Ddc* gene expression during development; during midpupation, titers of β-ecdysone again increase to near maximal levels [Kraminsky et al., 1980], but *Ddc* enzyme levels do not. Furthermore, the final peak of *Ddc* enzyme activity at adult eclosion occurs at a time when β-ecdysone titers are near basal levels and have been there for some 36 hr. Similarly, there is a peak of β-ecdysone titer at 10 hr embryonic development [Kraminsky et al., 1980] with no concurrent increase in levels of either *Ddc* enzyme activity or appearance of *Ddc* transcripts [Beall and Hirsh, 1984].

Levels of the *Ddc* mRNA are present at relatively high levels only during developmental periods when *Ddc* enzyme activity is subject to sharp increases [Hirsh and Davidson, 1981]. This demonstrates that the primary *Ddc* gene regulation occurs pretranslationally. This has been found also during embryonic development, over a shorter interval with higher temporal resolution [Beall and Hirsh, 1984]. *Ddc* enzyme activity is induced at 16–18 hr of embryonic development, reaching fully induced levels within 4 hr and a plateau at these levels for about 10 hr. *Ddc* transcripts first appear concurrently with the appearance of *Ddc* enzyme activity and are substantially reduced 4 hr later when enzyme levels are maximal. Complexity in *Ddc* gene expression is revealed by the observation of a number of transcripts from *Ddc* at this stage. Each of these species is larger than the mature *Ddc* mRNA and contains some or all of the *Ddc* intervening sequences. Only one species is uniquely found at this stage of development. This RNA, of length about 2.3 kb, contains a portion of the first intervening sequence (Fig. 1). It is maximally abundant at about 22 hr of development, significantly later than the time of maximal abundance of the *Ddc* 2.1 kb mRNA [Beall and Hirsh, 1984]. Though this species has 5' and 3' ends indistinguishable from the 2.1 kb RNA (Morgan and Hirsh, in preparation), we do not know whether this species is an alternately spliced or a partially spliced RNA.

A somewhat more complicated picture of *Ddc* expression during embryonic development has recently appeared [Gietz and Hodgetts, 1985]. This work indicates the presence of *Ddc* gene transcripts even during early embryogenesis before the onset of measureable Ddc enzyme activity, whereas Beall and Hirsh [1984] detected no *Ddc* transcripts before 16 hr embryonic development. Gietz and Hodgetts [1985] detected these early embryonic transcripts with a probe extending considerably beyond the 3' end of *Ddc*. We have found, however, that this downstream region encodes transcripts of 1.9 and 2.3 kb, nearly indistinguishable in length from the *Ddc* RNAs. These transcripts are expressed throughout much of embryonic development (Beall and Hirsh, in press).

Two observations indicate that posttranscriptional regulation must be involved in regulating *Ddc* gene expression during embryonic development [Beall and Hirsh, 1984]. First, intervening sequence containing RNAs are present at exceptionally high levels. During late embryonic development up to half of the *Ddc* RNA molecules can contain intervening sequences. This alone indicates that processing must be relatively slow relative to the mRNA half-life. Given the kinetics of mRNA decline during late embryogenesis, this half-life appears to be 2 hr or less. Second, as the 2.1 kb *Ddc* mRNA declines in abundance during late embryogenesis, the intervening sequence containing species disappear with a significantly longer half-life. This is not consistent with transcription being the rate-limiting step. It is not possible to define further which posttranscriptional step is rate limiting without knowledge of the cellular location of the different *Ddc* RNA species. Unfortunately, given the low abundance of the *Ddc* transcripts, this has so far proved impossible (Beall and Hirsh, unpublished results). More direct approaches, utilizing P element-mediated gene reintegration, are likely to be necessary to gain further insight into the nature of these limiting steps.

One mode of developmental regulation occasionally used when a particular gene must be expressed at multiple developmental times is to specify developmentally distinct RNA start points. This strategy is utilized by the *Drosophila* alcohol dehydrogenase gene *(Adh)* [Benyajati et al., 1983]. This strategy does not appear to be utilized by *Ddc*. An mRNA of the same length is induced at each developmental stage at which *Ddc* is expressed [Beall and Hirsh, 1984]. Higher-resolution studies employing primer extension mapping have located an identical RNA start point at each of these developmental stages (Morgan and Hirsh, in preparation).

D. P Element-Mediated Gene Reintegration as a Means of Studying Regulation

P elements are transposable elements naturally occuring in certain strains of *Drosophila*. As was first shown by Spradling and Rubin [1982], these elements can be used as vectors to achieve germline reintegration of cloned genes. Several *Drosophila* genes have been reintegrated using P elements, including the genes encoding the enzymes xanthine dehydrogenase (the *rosy* locus) [Spradling and Rubin, 1983], alcohol dehydrogenase (*Adh*) [Goldberg et al., 1983], dopa decarboxylase (*Ddc*) [Scholnick et al., 1983], and a salivary glue protein gene *sgs-3* [Richards et al., 1983]. The striking findings to emerge from these studies are that each of these genes is developmentally regulated when chromosomally reintegrated via P elements and that the degree of expression is only influenced to a small degree by chromosomal position. Examination of several strains carrying intact *Ddc* genes showed levels of *Ddc* expression within 35% of that found in wild type strains. Similar results were obtained when the levels of *rosy* gene expression were quantitated [Spradling and Rubin, 1983]. Thus this means of gene reintegration appears to be suitable for a detailed examination of the cis-acting sequence requirements for normal gene expression.

Our initial study showing regulated expression of *Ddc* utilized a DNA segment containing the *Ddc* gene plus 2.5 kb of 5' flanking sequences, and 1 kb of 3' flanking sequences [Scholnick et al., 1983]. The reintegrated genes are expressed with the same tissue and temporal specificity as their in situ wild type counterpart. We have since shown that levels of *Ddc* enzyme activity in the central nervous systems are restored to normal levels in transformed flies (unpublished data), and normal learning behavior is restored (Gailey, personal communication).

We have initiated a study of the cis-acting sequence requirements for developmental regulation of *Ddc*. Our first goal is localize the extent of the sequences necessary for normal *Ddc* function by reintegrating a set of DNA fragments containing decreasing amounts of 5' flanking sequences. These segments are illustrated in Figure 4. The *Ddc* segments were inserted into P element vectors containing a *Drosophila* alcohol dehydrogenase gene (*Adh*) as a selectable marker (modified versions of vectors obtained from J. Posakony). These vectors were reintegrated into a strain made doubly mutant for both *Adh* and *Ddc* genes such that detection of transformed flies was not dependent on *Ddc*⁺ function.

We have observed apparently normal developmental regulation with all *Ddc* genes containing 208 or more base pairs of 5' flanking DNA. Low-resolution developmental profiles of flies containing *Ddc* genes with 208 bp or 22 bp 5' flanking sequences are shown in Figure 5. *Ddc* genes containing only 22 bp of 5' flanking sequences do not show a normal developmental profile. *Ddc* expression is prematurely elevated during larval development but decreases after pupation and does not show normal induction at adult

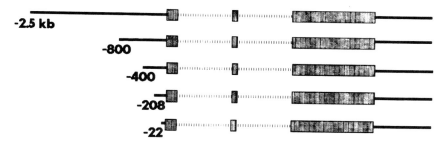

Fig. 4.   *Ddc* 5' deletions used to define sequences minimally necessary for developmentally regulated expression. Deletions were constructed by partially digesting with restriction enzymes, cutting at the appropriate locations, and converting these sites to Pst1 sites with linkers. The deleted genes were introduced to a P element vector containing the selectable gene *Adh*.

eclosion. The surprising finding is that these genes are expressed at all: The deletion has removed the normal "TATA" box sequences, and no similar sequences are found in the now adjacent polylinker sequences. We have recently demonstrated that the wild type RNA start site is used for expression of these mutant genes, at least during midlarval development (Morgan and Hirsh, in preparation). They are therefore not being expressed by read-through transcription from sequences from within the *Adh* gene. Thus some elements that normally determine the *Ddc* RNA start point must be retained in these constructs. We cannot eliminate the possibility that the observed expression is being influenced by sequences that are now juxtaposed to the *Ddc* RNA start point. Indeed, the high expression during larval development leads us to suspect that this *Ddc* expression is being influenced by some elements from within the adjacent *Adh* gene, which is normally expressed during this period.

III. PROSPECTS FOR FUTURE STUDIES

The *Ddc* gene is subject to striking regulation through development. Studies to date indicate that this regulation involves posttranscriptional as well as transcriptional steps. P element-mediated transformation of *Drosophila* has the potential for shedding light on the cis-acting sequence requirements for this regulated expression. Our preliminary experiments presented here indicate that cis-acting sequence elements very near or within the gene must participate in specifying temporally and tissue-specific regulation. In particular, sequences farther than 208 bp from the RNA start point do not appear to be essential for normal postembryonic temporal regulation. Sur-

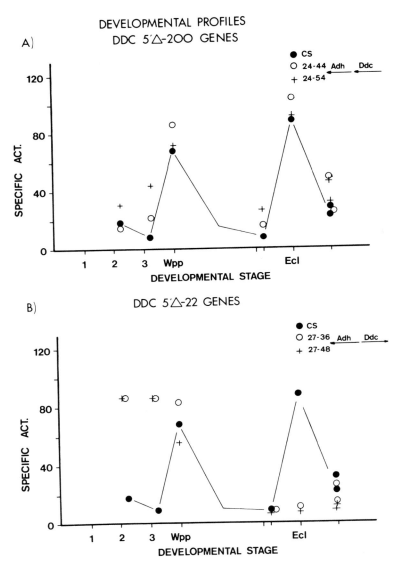

Fig. 5. Postembryonic developmental profiles of 5'-deleted *Ddc* genes reintegrated via P element vectors. The orientations of the *Adh* and *Ddc* genes in the P element vectors are shown, the arrows showing the directions of transcription. The lines show the expected Canton-S regulation during pupal development. 1, 2, 3, larval instar stages; Wpp, white prepupae; Ecl, newly eclosed adults.

prisingly, deleting all but 22 bp of 5' flanking sequences does not eliminate expression of the gene but instead leads to expression that at some developmental time points can be at near-normal levels. In that the expression observed with this construct is not normal and might be affected by the sequences newly juxtaposed at its 5' end, additional constructs that vary both the orientation within the P element vector and the nature of the 5' flanking sequences must be tested before firm conclusions can be drawn regarding the nature of the regulatory elements retained by the gene.

Where can these studies lead? If separable elements are indeed responsible for various facets of the tissue- and temporally regulated expression of *Ddc*, these elements could potentially be utilized in a number of ways. Being able to express selectively *Ddc* in one particular tissue could lead to a better understanding of the physiological effects of loss of *Ddc* expression. For instance, it has been observed that flies containing *Ddc* temperature-sensitive mutations do not learn normally at the restrictive temperature [Tempel et al., 1984]. This is most simply interpreted as an effect of loss of CNS function. However, this loss of normal learning ability could alternatively be owing to overall lowering of viability of these flies. A *Ddc* gene engineered with a selective loss of function in the central nervous system would allow a more precise interpretation of the behavioral effects.

Similarly, the existence of separable elements would make it feasible to consider altering the expression of other genes with *Ddc* regulatory elements. In this manner it might be possible to alter the physiology of specific tissues if not specific cells in either the central nervous system or the hypoderm.

ACKNOWLEDGMENTS

The work performed in this laboratory was supported by grants from the National Institutes of Health and the National Foundation—March of Dimes. The author thanks the various members of this laboratory for allowing communcation of unpublished results and for helpful comments on the manuscript.

IV. REFERENCES

Beall C, Hirsh J (1984): High levels of intron containing RNAs are associated with expression of the *Drosophila* dopa decarboxylase gene. Mol Cell Biol 4:1669–1674.

Benyajati C, Spoerel N, Haymerle H, Ashburner M (1983): The messenger RNA for alcohol dehydrogenase in *Drosophila melanogaster* differs in its 5' end in different developmental stages. Cell 33:125–133.

Brunet PCJ (1980): The metabolism of the aromatic amino acids concerned in the cross-linking of insect cuticle. Insect Biochem 10:467–500.

Clark WC, Pass PS, Bhagyalakshmi V, Hodgetts RB (1978): Dopa decarboxylase from *Drosophila melanogaster*. Mol Gen Genet 162:287–297.

Dewhurst SA, Croker SG, Ikeda K, McCaman RE (1972): Metabolism of biogenic amines in *Drosophila* nervous tissue. Comp Biochem Physiol 43B:975–981.

Gietz RD, Hodgetts RB (1985): An analysis of dopa decarboxylase expression during embryogenesis in *Drosophila melanogaster*. Dev Biol 107:142–155.

Gilbert D, Hirsh J, Wright TRF (1984): Molecular characterization of a gene cluster flanking the *Drosophila* dopa decarboxylase gene. Genetics 106:679–694.

Goldberg D, Posakony J, Maniatis T (1983): Correct developmental expression of a cloned alcohol dehydrogenase gene transduced into the *Drosophila* germ line. Cell 34:50–73.

Hirsh J, Davidson N (1981): Isolation and characterization of the *Drosophila* dopa decarboxylase gene. Mol Cell Biol 1:475–485.

Hodgetts RB (1975): The response of dopa decarboxylase activity to variations of gene dosage in *Drosophila*: A possible location of the structural gene. Genetics 79:45–54.

Kraminsky GP, Clark WC, Estelle MA, Gietz RD, Sage BA, O'Connor JD, Hodgetts RB (1980): Induction of translatable mRNA for Dopa decarboxylase in *Drosophila*: An early response to ecdysterone. Proc Natl Acad Sci USA 77:4175–4179.

Livingstone MS, Tempel BL (1983): Genetic dissection of monamine neurotransmitter synthesis in *Drosophila*. Nature 303:67–70.

Marsh JL, Wright TRF (1980): Developmental relationship between dopa decarboxylase, dopa acetyltransferase, and ecdysone in *Drosophila*. Dev Biol 80:379–387.

Richards G, Cassab A, Bourouis M, Jarry B, Dissous C (1983): The normal developmental regulation of a cloned sgs3 'glue' gene chromosomally integrated in *Drosophila melanogaster* by P element transformation. EMBO J 2:2137–2142.

Scholnick S, Morgan BA, Hirsh J (1983): The cloned dopa decarboxylase gene is developmentally regulated when reintegrated into the Drosophila germline. Cell 34:37–45.

Spradling AC, Rubin GM (1982): Transposition of cloned P elements into *Drosophila* germline chromosomes. Science 218:341–347.

Spradling AC, Rubin GM (1983): The effect of chromosomal position on the expression of the *Drosophila* xanthine dehydrogenase gene. Cell 34:47–57.

Tempel BL, Livingston MS, Quinn WG (1984): Mutations in the dopa decarboxylase gene affect learning in *Drosophila*. Proc Natl Acad Sci USA 81:3577–3581.

White K, Valles AM (1984): Immunohistochemical and Genetic Studies of Serotonin and Neuropeptides in *Drosophila*. In Edelman BM, Gall E, Cowan WM (eds): "Molecular Basis of Neural Development." New York: John Wiley and Sons, pp 547–564.

Wright TRF, Bewley GC, Sherald AF (1976a): The genetics of dopa decarboxylase in *Drosophila melanogaster* II. Isolation and characterization of dopa decarboxylase deficient mutants and their relationship to the alpha methyl dopa hypersensitive mutants. Genetics 84:287–310.

Wright TRF, Black BC, Bishop CP, Marsh JL, Pentz ES, Steward R, Wright EY (1982): The genetics of dopa decarboxylase in *Drosophila melanogaster* V. Ddc and *1(21)amd* alleles: Isolation, characterization and intragenic complementation. Mol Gen Genet 188:18–26.

Wright TRF, Hodgetts RB, Sherald AF (1976b): The genetics of dopa decarboxylase in *Drosophila melanogaster* I. Isolation and characterization of deficiencies that delete the dopa decarboxylase dosage sensitive region and alpha methyl dopa hypersensitive locus. Genetics 84:267–285.

IV. Foreign Genes in Eukaryotes

Molecular Developmental Biology, pages 117–130
© 1986 Alan R. Liss, Inc.

Stimulation of In Vitro Transcription by the SV40 Enhancer Involves a Specific *Trans*-Acting Factor

Pierre Chambon, Alan Wildeman, and Paolo Sassone-Corsi

Laboratoire de Génétique Moléculaire des Eucaryotes du CNRS, Unité 184 de Biologie Moléculaire et de Génie Génétique de l'INSERM, Faculté de Médecine, 67085 Strasbourg, France

I. INTRODUCTION

Regulation of gene expression at the level of transcription is most likely an important control mechanism during development and in the terminally differentiated cells of eukaryotic organisms. It is commonly though that this control results from the interaction between specific DNA sequences, regulatory proteins, and the transcriptional machinery [Darnell, 1982]. Recent studies have shown that the promoter DNA sequences involved in the control of transcription initiation of eukaryotic protein-coding genes are composed of several elements: the mRNA startsite, the TATA sequence, and one or several upstream elements generally located within approximately 110 bp upstream of the capsite [Breathnach and Chambon, 1981; McKnight et al., 1984]. For some transcription units, additional elements called *enhancers*, which were discovered in Simian Virus 40 (SV40) [Benoist and Chambon, 1981; Gruss et al., 1981] and in other viruses [for reviews see Yaniv, 1982;

Khoury and Gruss, 1983], and more recently were found in cellular genes [Banerji et al., 1983; Gillies et al., 1983; Queen and Baltimore, 1983], are important for efficient transcription in vivo.

Enhancers are *cis*-acting elements increasing dramatically the transcription from those genes where they occur, acting on their natural promoter or heterologous promoters [Moreau et al., 1981; Wasylk et al., 1983] in either orientation and over distances of several kb pairs (kbp). They may also play a role in cell or tissue specificity of gene expression [see Yaniv, 1982; Khoury and Gruss, 1983; Banerji et al., 1983; Gillies et al., 1983; Queen and Baltimore, 1983; Chandler et al., 1983], and there has been some suggestion that specific transription factors recognize these sequences [Schöler and Gruss, 1984].

The SV40 enhancer, which contains the 72 bp repeat, has been extensively studied [Moreau et al., 1981; Fromm and Berg, 1982; Wasylk et al., 1983], and many of its characteristics are common to other enhancer sequences. Furthermore, as has been demonstrated using DNaseI hypersensitivity assays and electron microscopy, the SV40 enhancer induces an alteration in chromatin structure over its own sequence [Jongstra et al., 1984]. In vitro transcription systems have demonstrated that for a number of promoter regions specific cellular factors interact with the TATA box element [e.g., Davison et al., 1983; Parker and Topol, 1984a] and upstream sequence elements [Dynan and Tjian, 1983; Parker and Topol, 1984b]. Recently, using whole cellular extracts (WCE), we have shown that the SV40 enhancer can stimulate specific in vitro transcription from heterologous promoter elements [Sassone-Corsi et al., 1984]. Moreover, deletion mutations that strongly decrease enhancer activity in vivo also abolished the in vitro stimulation. In that the template does not appear to be organized into a chromatin structure during the in vitro synthesis [Hen et al., 1982], this result suggested to us that the observed stimulation was due to the interaction between some transcriptional factor(s) and the enhancer sequence. We now demonstrate that in nuclear extracts (NE) the homologous SV40 early promoter is efficiently transcribed and that under optimal ionic conditions and DNA concentrations this transcription is decreased 10- to 15-fold by deletion of the enhancer. This effect is observed even when the enhancer and the upstream sequence element (the 21 bp repeat) are not in close apposition. In addition, we report here results obtained from in vitro competition experiments using both WCE and NE indicating that a *trans*-acting factor(s) is responsible for the SV40 enhancer activity in vitro. We also show that the "enhancer factor" is different from those factors that interact with the upstream elements of the adenovirus-2 major late (Ad2ML) or SV40 early (21 bp repeat region) promoters. It appears that a stable complex is formed between the enhancer and the *trans-*acting factor(s) and that sequences located both in the 3′ and 5′ regions of the 72 bp repeat are involved in the formation of this complex.

II. IN VITRO ANALYSIS OF ENHANCER FUNCTION

A. Stimulation of In Vitro Transcription by the SV40 Enhancer Is Due to a *Trans*-Acting Factor(s)

Analysis of rabbit β-globin RNA synthesized from plasmids pAO and pA56 [recombinants having an SV40 early promoter region with an enhancer containing a single 72 bp sequence or with no enhancer, respectively (see Fig. 1) after calcium-phosphate transfection into HeLa cells has revealed that the presence of the enhancer results in a dramatic stimulation of transcription from the SV40 early mRNA start sites (Zenke et al., in preparation). Using HeLa whole cell [Manley et al., 1980] or S100 [Weil et al., 1979] extracts the in vivo effect of the enhancer is reproduced only slightly, with a two- to four-fold stimulation of transcription normally being observed. Nuclear extracts [Dignam et al., 1983], however, enable a clean run-off analysis of transcription from the SV40 early promoter, and by using optimal concentrations of $MgCl_2$, spermidine, and DNA template a deletion of the enhancer results in a ten- to fifteen-fold decrease in transcription [Wildeman et al., 1984].

A competition assay was used to investigate whether this stimulation of transcription by the 72 bp repeat could be owing to a specific *trans*-acting factor(s). In vitro transcriptions of AccI-digested pAO and pA56 (see Fig. 1) were carried out in the presence of various purified promoter elements to monitor the competition for transcription factors. The initial experiments were designed to see if transcription from the SV40 early promoter could be selectively competed for by a DNA fragment containing both the 21 bp repeat region and the enhancer. A purified KpnI-NcoI fragment (Fig. 1) spanning this region was found to compete efficiently for transcription of pAO (Fig. 2, lanes 2–4), whereas a fragment containing the upstream element of the Ad2MLP did not (Fig. 2, lanes 5–7). This result is in agreement with previous observations [Miyamoto et al., 1984] that showed that the upstream element of the Ad2MLP did not compete out the effect of the 21 bp repeat region. When increasing amounts of a fragment containing the enhancer (KpnI-BamHI segment of pAO, see Fig. 1) were used as a competitor, transcription of pAO template was reduced to the level of that seen from a pA56 template (Fig. 2B). Such a result would be expected only if this fragment were trapping a factor(s) that acts on the enhancer present in the template DNA.

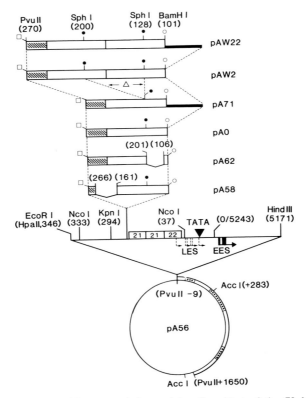

Fig. 1. Structure of recombinants used for studying the effect of the 72 bp repeat on transcription from the SV40 early promoter. pA56 contains the SV40 origin from HpaII (346) to HindIII (5171) fused to a rabbit β-globin coding sequence from −9 to +1650 [Van Ooyen, 1979] (double line), and pBR322 sequences (single line) from PvuII (2066) to EcoRI (4361). Within the SV40 sequences, a deletion extending from a BamHI site (101) generated by in vitro site-directed mutagenesis of nucleotides TAGTCC (106–101) to GGATCC (Grundström et al., in preparation) to the PvuII site (270) created an enhancerless promoter. pAW2 and pAW22 contain a wild type 72 bp repeat region with pAW22 having a pBR322 fragment (BamHI to NarI, coordinates 375–413, with Bgl II linker ligated to the NarI site) inserted at the BamHI site (solid line). pAO and pA71 have a single 72 bp sequence [generated by a deletion between SphI (128) and SphI (200)] with pA71 carrying the same pBR322 insertions as pAW22. pA62 and pA58 were derived from pAO. pA62 contains a deletion from 106 to 201 generated by a cutting at BamHI and SphI sites, followed by blunt-ending with Klenow enzyme and litigation. The deletion present in pA58 has been described previously [mutant TB101 in Moreau et al., 1981], and was cloned into the pA series by a KpnI (294)–SphI (128) transfer. Early-early (EES) and late-early (LES) start sites and TATA box and 21 base pair repeat regions are indicated [for references see Vigneron et al., 1984]. Hatched areas are sequences between the 5' end of the 72 bp repeat (coordinate 251) and the PvuII site (270). Nucleotide coordinates for SV40 follow the BBB system [Tooze, 1982].

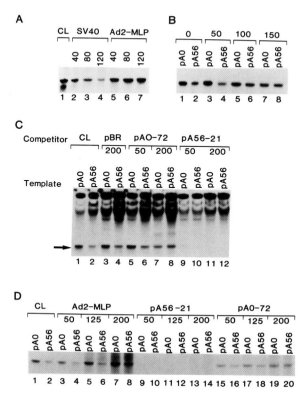

Fig. 2. Transcription from the SV40 early promoter is stimulated by an enhancer-specific *trans*-acting factor(s). Either pAO or pA56 template (100 ng) was transcribed in the presence of various purified competitor fragments as described by Wildeman et al. [1984] using MgCl₂, spermidine, and KCl concentrations of 4.5 mM, 3.0 mM, and 50 mM, respectively. The run-off transcripts are those originating from the EES. Long exposures of these gels, and S1 nuclease mapping of these transcripts, reveal both the EES and LES present in vivo. A, transcription of pAO in the presence of no competitor (CL), and 40, 80, or 120 ng of an NcoI-NcoI fragment (SV40 coordinates 37–333) of pAO containing the SV40 21 bp and 72 bp repeat promoter elements (SV40) or a XhoI-XhoI fragment from the −260 to −29 positions of the Ad2MLP [Hen et al., 1982]. The XhoI site at −29 was created in vitro by site-directed mutagenesis (a gift of R. Hen and M. Wintzerith). B, transcription of pAO and pA56 in the presence of 0, 50, 100, or 150 ng of a BamHI-KpnI fragment (SV40 coordinates 101 and 294, respectively) from pAO. C, transcription of pAO and pA56 in the presence of no competitor (CL), 200 ng of an EcoRV to BamHI fragment of pBR322 (position 185–375; pBR), 50 and 200 ng of the BamHI-KpnI fragment used in B (pAO-72), or 50 and 200 ng of a NcoI-KpnI fragment used in B (pAO-72), or 50 and 200 ng of a NcoI-KpnI fragment (coordinates 37 and 294) and pA56 containing the 21 bp repeat (pA56-21). D, transcription of pAO and pA56 in the presence of no competitor (CL) or 50, 125, and 200 ng of the Ad2MLP XhoI-XhoI competitor fragment (Ad2MLP), the 21 bp repeat fragment (pA56-21), or the 72 bp fragment (pAO-72).

In Figures 2A and B it can be seen that the level of transcription of pAO and of pA56 rises slightly as more Ad2MLP or pAO fragment, respectively, are added. This phenomenon is also seen when nonspecific DNA, such as pBR322, is added to the in vitro transcription system [e.g., Dignam et al., 1983]. To ensure that the competition effects observed were indeed enhancer-specific and not owing to changes in DNA concentrations, various competitors were tested simultaneously using the same extract and DNA template preparations. Figure 2C shows that a pBR322-derived fragment (lanes 3 and 4) did not compete for transcription from either pAO or pA56. The fragment containing an enhancer sequence (lanes 5–8) competed only for the transcription from a promoter with an enhancer (pAO) but not for that from an enhancerless template (pA56). At the same time, however, a fragment containing only the upstream element (the 21 bp repeat, lanes 9–12) competed for the transcription from either template, indicating that the enhancer factor is not acting on the upstream element. Figure 2D shows that, when both the enhancer factor(s) and upstream sequence factor(s) were effectively competed for by their respective elements, the Ad2MLP upstream sequence again was unable to compete for transcription from either SV40 template.

In all the competition experiments, it was found that the amount of 72 bp repeat competitor required to compete totally for the effect of the enhancer ($\sim$ 150–200 ng) represents a 20- to 40-fold molar excess. With respect to the template DNA, this excess is about five times greater than that required to compete for 21 bp factor(s). Whether these molar values reflect a greater quantity of available enhancer-binding factor(s) than upstream sequence binding factor(s) in the extract or a difference in the stabilities of the complexes between the competitor fragments and their cognate factors should become more apparent as the nuclear extract is further purified. It is interesting to note that, as is shown in Figure 2C, the difference in background transcription between pAO and pA56 templates, visible at the top of the gel, is immediately lost as 21 bp repeat sequences begin to compete. This suggests that the Sp1 21 bp repeat transcription factor [Dynan and Tjian, 1983] is activating background transcription and that the presence of the enhancer enables this factor to be more selective.

B. The 5′ and 3′ Domains of the Enhancer Are Both Required for Binding of the Factor

Constructions pA'58 and pA62 (see Fig. 1) contain internal deletions in the pAO enhancer that reduce its efficiency in vivo by 50- and 15-fold, respectively (Zenke et al., in preparation). Competitor fragments (KpnI-BamHI segments) were prepared from these plasmids and tested for their ability to compete for enhancer-dependent transcription using pAW22 and pA56 as templates. As is shown in Figure 3, a competitor fragment derived from pAO and containing an intact enhancer (lanes 3–6) competes only for transcription from the pAW22 template. When fragments derived from pA58 (lanes 7–10) and pA62 (lanes 11–14) were used, the efficiency of competition was strongly reduced compared to that of the fragment from pAO. Some weak competition against the pAW22 template was observed with 200 ng of these competitors lacking either the 5' or the 3' domain of the enhancer (lanes 9 and 13). These results show that a single 72 bp sequence can compete with an enhancer containing a 72 bp repeat and that the competition is unaffected when the enhancer and 21 bp repeat elements are separated.

C. Transcriptional Factors Interacting With Upstream Promoter Elements and Enhancers Are Different

The competition experiments using the homologous promoter show that the SV40 enhancer factor is different from the factor recognizing the Ad2ML promoter upstream sequence. To investigate further whether the SV40 21 bp repeat region and the Ad2MLP upstream element could bind the enhancer factor, heterologous constructions were used.

We have previously reported that using a HeLa WCE the SV40 enhancer stimulates transcription in vitro in *cis* from heterologous promoter elements irrespective of its orientation [Sassone-Corsi et al., 1984]. Inserting the SV40 72 bp repeat in close apposition to the −34/+33 Ad2ML promoter (Ad2MLP) element (recombinant pSVBA34, see Fig. 4A) resulted in a ten-fold stimulation of specific in vitro transcription, and deletions known to diminish the enhancer function in vivo also decreased its activity in vitro. As with the homologous promoter, a competition assay reveals that this in vitro

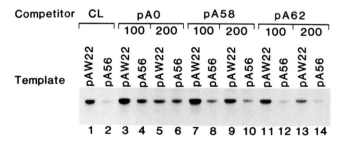

Fig. 3.   Mutations in the SV40 enhancer diminish its ability to trap a *trans*-acting factor(s). Transcriptions were carried out as in Figure 2, using 200 ng of pAW22 or pA56 template, in the presence of no competitor (CL), or 100 and 200 ng of BamHI-KpnI fragments prepared from pAO, pA58, or pA62 (see Fig. 1).

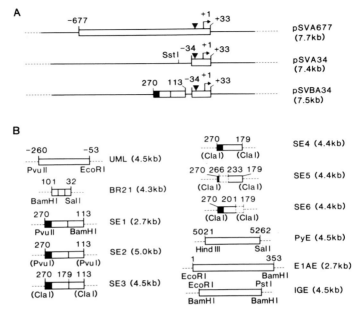

Fig. 4.   Structure of the heterologous recombinants used as templates or as competitors in the in vitro transcription analysis. A, the hybrid template constructions pSVA677, pSVA34, and pSVBA34 have been described [Hen et al., 1982; Sassone-Corsi et al., 1984]. pSVA677 contains the entire adenovirus-2 major late promoter (Ad2MLP) between coordinates −677 and +33 with respect to the capsite. pSVA34 contains a deletion [Corden et al., 1980] that removes the upstream promoter sequences between −677 and −34 but leaves the TATA box (▼) intact. pSVBA34 is derived from pSVA34 by inserting an SV40 fragment, containing the 72 bp repeat (double open box, coordinates 113–251, BBB system) [Tooze, 1982] and an additional 5′ flanking sequence (■) up to coordinate 270, into an SstI site located 63 bp upstream from the Ad2MLP capsite [Hen et al., 1982]. TaqI digests of these recombinants were used as templates in a HeLa WCE in vitro transcription system, generating a specific run-off of 525 nucleotides [Sassone-Corsi et al., 1984]. B, structure of the plasmids used as competitors in the in vitro prebinding transcription experiments. DNA fragments containing isolated enhancer sequences or upstream promoter elements were cloned into the pBR322 [Sutcliffe, 1978] sites as indicated. UML contains the Ad2MLP upstream sequences between positions −53 and −260 [Hen et al., 1982] inserted between the PvuII and the EcoRI sites. BR21 contains the SV40 early promoter upstream element (21 bp repeat region) [Baty et al., 1983] between coordinates 32 and 101 inserted between the SalI and BamHI sites. SE1 contains the SV40 enhancer from coordinates 113 to 270 [Moreau et al., 1981] inserted between the BamHI and PvuII sites; SE2 contains the same enhancer inserted into the PvuI site; SE3, SE4, SE5, SE6, PyE, E1AE, and IGE competitors are described and their effects analyzed by Sassone-Corsi et al. [1985] and are not used in the present report.

stimulation is owing to the presence of a trans-acting factor(s) [Sassone-Corsi et al., 1985]. A HeLa WCE was first preincubated with increasing amounts of competitor DNA containing isolated promoter or enhancer elements (see Fig. 4B). After 15 min at 25°C, the template and nucleoside triphosphates were added. Preincubation of the WCE with increasing molar amounts of SE1 [a pBR322 plasmid containing only the SV40 enhancer (Fig. 4B)] decreased specific transcription from the −34/+33 Ad2MLP element only when it was linked to the SV40 enhancer (compare pSVBA34 and pSVA34 templates in Fig. 5A, lanes 1–5, and B, lanes 1–4). Transcription from the "competed" pSVBA 34 was similar to that from pSVA34. These results confirm that stimulation of in vitro transcription by the enhancer involves a trans-acting factor(s). Although lower molar ratios of competitor to template were required for efficient competition in WCE compared to nuclear extract, we do not know at present if this is because of different concentrations of the enhancer factor in each extract or because of differences in the assay conditions.

To investigate whether or not the enhancer factor is different from factors that interact with upstream promoter elements, preincubations were carried

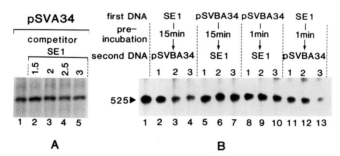

Fig. 5. A transcriptional factor interacts with the SV40 enhancer. A, in vitro transcription run-off analysis of the competition effect of SE1 on pSVA34 transcription. Lane 1, transcription of the pSVA34 template alone; lanes 2–5, effect of preincubating the WCE for 15 min at 25°C with increasing molar amounts of SE1 competitor DNA (1.5–3 molar excess of SE1 to pSVA34). B, effect of SE1 competition on the in vitro transcription of pSVBA34. Lane 1, transcription of the pSVBA34 template alone; lanes 2–4, WCE was preincubated 15 min with increasing amounts of SE1 (molar ratio vs. pSVBA34, 1–3) prior to the addition of pSVBA34 template; lanes 5–7, preincubation of the WCE with pSVBA34 for 15 min and subsequent addition of the SE1 competitor DNA; lanes 8–10, as in lanes 5–7 but the preincubation time was 1 min; lanes 11–13, as in lanes 2–4 but the preincubation time was 1 min. The numbers above the electrophoretic gel lanes indicate the molar ratio of competitor to template DNA. The size of the run-off RNA is 525 nucleotides. The in vitro transcription assay and the preparation of HeLa WCE have been described [Manley et al., 1980; Sassone-Corsi et al., 1981]. For a description of the competition assay, see Sassone-Corsi et al. [1985].

out with plasmids containing either the SV40 early (21 bp repeat) or the Ad2MLP upstream elements. It has been shown that the Sp1 factor [Dynan and Tjian, 1983], which interacts with the 21 bp repeat, is different from that interacting with the upstream element of the Ad2MLP [Dynan and Tjian, 1983; Miyamoto et al., 1984]. Specific transcription from pSVA677 (see Fig. 4A), a plasmid that contains the intact Ad2MLP [Sassone-Corsi et al., 1984], was decreased by preincubation with the recombinant (UML), which contains the Ad2MLP upstream sequences between −53 and −260 (Fig. 6A, lanes

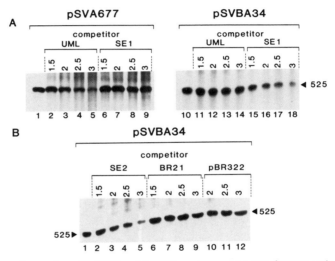

Fig. 6.   Factors interacting with SV40 and Ad2ML upstream promoter elements and enhancer are different. A, run-off analysis of the effect of competition by UML and SE1 on transcription from pSVA677 (lanes 1–9) and pSVBA34 (lanes 10–18). Lane 1, pSVA677 template transcription in the absence of competitor DNA; lanes 2–5, competition effect of UML as its concentration is increased in the preincubation (1.5–3 molar ratio); lanes 6–9, effect of preincubating SE1 at increasing molar ratios; lane 10, RNA synthesized with pSVBA34 template when no competitor DNA is added; lanes 11–14, absence of competition effect by UML at molar ratios increasing from 1.5 to 3; lanes 15–18, increasing competition effect of SE1 on pSVBA34 transcription using the same range of molar ratios. B, run-off analysis of the effects of SE2, BR21, or pBR322 competitor DNA on pSVBA34 transcription. Lane 1, pSVBA34 transcription in the absence of competitor; lanes 2–5, competition when SE2 is preincubated with the WCE at molar ratios increasing from 1.5 to 3; lanes 6–9, increasing amounts of BR21 were used for the preincubation; lanes 10–12, effect of preincubating the WCE with pBR322 DNA prior the addition of the pSVBA34 template. The numbers above the electrophoretic gel lanes indicate the molar ratio of competitor to template DNA. The size of the run-off RNA is 525 nucleotides. Methods are as previously described [Manley et al., 1980; Sassone-Corsi et al., 1981; Sassone-Corsi et al., 1985].

2–5), but not by preincubation with SE1 even at the highest template/ competitor molar ratios (lanes 6–9). The converse experiment, using pSVBA34 template, gave the expected results. Transcription from pSVBA34 was inhibited by preincubation with SE1 (lanes 15–18) but not with UML (lanes 11–14).

The above results indicate that the factor interacting with the upstream region of the Ad2MLP is different from that interacting with the SV40 enhancer. They also demonstrate that the in vitro stimulatory effect of the SV40 enhancer, using the Ad2MLP hybrid constructions (Figs. 4A and 5) [Sassone-Corsi et al., 1984], is not owing to the interaction between the 72 bp repeat and a factor(s) that binds to the Ad2MLP upstream promoter element.

To investigate whether or not the SV40 early promoter region (the 21 bp repeat) can bind the enhancer factor, competition experiments were performed using pSVBA34 as template and a pBR322-derived plasmid containing the 21 bp repeat sequence as competitor (BR21, see Fig. 4B). Lanes 2–5 in Figure 6B show the expected competition with the enhancer-containing plasmid SE2 (similar to SE1 but with the SV40 enhancer inserted elsewhere in pBR322, see Fig. 4B). Neither the 21 bp repeat element (plasmid pBR21, lanes 6–9) nor pBR322 alone (lanes 10–12) could compete for transcription from pSVBA34. Thus the SV40 Sp1 factor does not appear to be responsible for the in vitro stimulation brought about by the SV40 enhancer.

The other competitors described in Figure 4B were also used to determine whether or not with heterologous constructions both 3' and 5' domains of the SV40 enhancer are necessary for the binding of the factor. The results, analogous to those obtained with the homologous promoter (see above), are described elsewhere [Sassone-Corsi et al., 1985]. The competitors PyE, E1AE, and IGE (see Fig. 4) contain enhancer elements from polyoma virus, adenovirus-2, and a heavy-chain immunoglobulin gene, respectively. The results obtained using these competitors are also described by Sassone-Corsi et al. [1985].

D. The Binding of the SV40 Enhancer Factor to the 72 bp Repeat Is Stable and Rapid

The stability of interaction of the enhancer factor with the SV40 72 bp repeat was analyzed in a prebinding experiment in which the pSVBA34 template was preincubated with the WCE (Fig. 5B, lanes 5–7). The second DNA, SE1, was added in increasing molar amounts after 15 min. No competition was observed even when a three fold molar excess of SE1 competitor DNA was added. When the time of prebinding was reduced to 1 min (Fig. 2B, lanes 8–13) no competition was visible when pSVBA34 was preincubated (lanes 8–10), whereas competition could be seen when the SE1 competitor DNA was present during preincubation (lanes 11–13). We conclude from these results that the in vitro stimulation by the SV40 enhancer involves the rapid formation of a stable enhancer-factor complex. That transcription from the pSVA34 template was not inhibited by preincubation with the SE1 enhancer-containing plasmid (Fig. 5A) also strongly suggests that the factor that interacts with the enhancer is not required for transcription from a template containing only the TATA box region.

III. DISCUSSION AND PERSPECTIVES

We have previously shown that the SV40 enhancer can stimulate transcription from heterologous promoter elements in vitro [Sassone-Corsi et al., 1984]. Using nuclear extracts, a similar stimulation can be observed with the homologous promoter elements present. The competition experiments demonstrate that transcription of the SV40 early promoter in vitro is stimulated by a specific *trans*-acting factor(s) that acts on the enhancer element. This factor appears to be different from those that act on upstream promoter elements, and mutations that both in vivo and in vitro diminish the ability of the enhancer to stimulate transcription also diminish its ability to compete for the factor in a transcription reaction. It seems clear, therefore, that the mechanism of enhancer function involves at least in part the specific recognition of these promoter elements by cellular factors.

At present it is not possible to say whether other properties of enhancers contribute to their function. The SV40 enhancer is capable of creating an "open" chromatin structure in vivo [Jongstra et al., 1984, and references therein], but whether this is owing to specific proteins that are bound to the enhancer or because of some intrinsic property of the sequence [Wasylyk et al., 1979] is not known. If it is owing to specific proteins, it is possible that the factor(s) revealed in the present in vitro study is also involved in the generation of the open chromatin structure. The availability of an in vitro system will now enable both a further purification of the factor and a closer analysis of its interaction with the DNA template. The fact that an enhancer fragment of only 120 nucleotides is able to trap a factor suggests that chromatin structure is not necessary for recognition of the enhancer by the factor, and studies to examine the structure of the competitor fragment in the transcription reaction are currently in progress. This factor(s) is clearly different from those factors that stimulate the transcription by binding to the TATA box [Davison et al., 1983; Parker and Topol, 1984a] or the upstream elements of the Ad2ML and SV40 early promoters [Hen et al., 1982; Dynan and Tjian, 1983; Miyamoto et al., 1984] (see Fig. 6). Furthermore, under the present in vitro conditions the enhancer sequence and the factor(s) can form a stable complex. The same 5′ and 3′ domains that are critical for enhancer activity in vivo are required in vitro for formation of this complex (see Fig. 3), implying a relatively large protein contact site, the binding of more than one protein, or a coiled enhancer sequence. It is worth noting that both the activity of the enhancer in vitro and its ability to form the stable complex are observed with a linear template and thus do not require super-helicity. Preliminary experiments to compete with a mixture of the two fragments that each carry a deletion (i.e., prepared from pA58 and pA62) have been unsuccessful, suggesting that the two deletions do not represent binding sites for two different proteins or that, if they do, these proteins must interact with each other to activate the promoter.

The binding of an enhancer-specific factor, demonstrated in the present study, to its cognate element appears to have a stability in vitro similar to that of other factors (our unpublished results). Considering that the upstream element (the 21 bp repeat region) [for references see Baty et al., 1983; Vigneron et al., 1984] and the enhancer [Moreau et al., 1981; Sassone-Corsi et al., 1984; our unpublished results] of the SV40 early promoter can still function when their position and/or orientation with respect to the capsite and the TATA box element are modified, it is interesting to speculate how the factors that bind to these three promoter elements can cooperate to promote efficient transcription. There must be either bending of the DNA template or movement of the factors from their recognition site to bring them into the necessary proximity. Alternatively, binding of the factors to their recognition sites might result in transmission of some DNA perturbation along the template by an as yet unknown mechanism. Considering that the SV40 enhancer factor operates in vitro either when the enhancer is in its natural position or is moved away from the 21 bp repeat region (see Figs. 1 and 3), it should be possible to explore further the mechanism invoved. Because of the nature of the DNA template in vitro, it seems probable that DNA supercoiling is not required for the positional flexibility of promoter elements.

IV. REFERENCES

Banerji J, Olson L, Schaffner W (1983): Cell 33:729–740.
Baty D, Barrera-Saldana H, Everett R, Vigneron M, Chambon P (1983): Nucleic Acids Res 12:915–932.
Benoist C, Chambon P (1981): Nature 290:304–310.
Breathnach R, Chambon P (1981): Annu Rev Biochem 50:349–383.

Chandler VL, Maler BA, Yamamoto KR (1983): Cell 33:489–499.

Corden J, Wasylyk B, Buchwalder A, Sassone-Corsi P, Kédinger C, Chambon P (1980): Science 209:1406–1414.

Darnell JE Jr. (1982): Nature 297:365–371.

Davison BL, Egly JM, Mulvihill ER, Chambon P (1983): Nature 301:680–686.

Dignam JD, Lebovitz RM, Roeder RG (1983): Nucleic Acids Res 11:1475–1489.

Dynan WS, Tjian R (1983): Cell 35:79–87.

Fromm M, Berg P (1982): J Mol App Genet 1:457–481.

Gillies SD, Morrison SL, Oi VT, Tonegawa S (1983): Cell 33:717–728.

Gruss P, Dhar R, Khoury G (1981): Proc Natl Acad Sci USA 78:943–947.

Hen R, Sassone-Corsi P, Corden J, Gaub MP, Chambon P (1982): Proc Natl Acad Sci USA 79:7132–7136.

Khoury G, Gruss P (1983): Cell 33:313–314.

Jongstra J, Reudelhuber TL, Oudet P, Benoist C, Chae CB, Jeltsch JM, Mathis DJ, Chambon P (1984): Nature 307:708–714.

Manley JL, Fire A, Cano A, Sharp PA, Gefter ML (1980): Proc Natl Acad Sci USA 77:3855–3859.

McKnight SL, Kingsbury RC, Spence A, Smith M (1984): Cell 37:253–262.

Miyamoto NG, Moncollin V, Wintzerith M, Hen R, Egly JM, Chambon P (1984): Nucleic Acids Res. 12:8779–8799.

Moreau P, Hen R, Wasylyk B, Everett RD, Gaub MP, Chambon P (1981): Nucleic Acids Res 9:6047–6068.

Parker C, Topol J (1984a): Cell 36:357–369.

Parker C, Topol J (1984b): Cell 37:273–283.

Queen C, Baltimore D (1983): Cell 33:741–748.

Sassone-Corsi P, Corden J, Kedinger C, Chambon P (1981): Nucleic Acids Res 9:3941–3958.

Sassone-Corsi P, Dougherty J, Wasylyk B, Chambon P (1984): Proc Natl Acad Sci USA 81:308–312.

Sassone-Corsi P, Wildeman A, Chambon P (1985): Nature 313:3129–3133.

Schöler HR, Gruss P (1984): Cell 36:403–411.

Sutcliffe JG (1978): Nucleic Acids Res 5:2721–2728.

Tooze J (ed) (1982): DNA Tumor Viruses. Cold Spring Harbor, New York: Cold Spring Harbor Laboratory.

Van Ooyen A, Van den Berg J, Mantei N, Weissman C (1979): Science 206:337–344.

Vigneron M, Barrera-Saldana H, Baty D, Everett R, Chambon P (1984): EMBO J (in press).

Wasylyk B, Oudet P, Chambon P (1979): Nucleic Acids Res 7:705–713.

Wasylyk B, Wasylyk C, Augereau P, Chambon P (1983): Cell 32:503–514.

Weil PA, Segall J, Harris B, Ng SY, Roeder RG (1979): J Biol Chem 254:6163–6173.

Wildeman A, Sassone-Corsi P, Grundström T, Zenke M, Chambon P (1984): EMBO J 3:3129–3133.

Yaniv M (1982): Nature 297:17–18.

Molecular Developmental Biology, pages 131–148
© 1986 Alan R. Liss, Inc.

Hormone-Dependent Transcriptional Enhancement and Its Implications for Mechanisms of Multifactor Gene Regulation

Keith R. Yamamoto

Department of Biochemistry and Biophysics, University of California, San Francisco, California 94143-0448

I. INTRODUCTION

Many developmental and physiological processes are triggered or modulated via the actions of small evolutionarily conserved "signalling molecules," such as steroid hormones. Steroids are virtually ubiquitous in eukaryotes, and their actions have been described in organisms ranging from simple fungi [Timberlake and Orr, 1984; Loose et al., 1981; Burshell et al., 1984] to primates [Anderson, 1984]. Steroid hormone effects are mediated by hormone-specific receptor proteins, which upon specific association with hormone ligands acquire increased affinity for chromosomal binding sites. One mode by which steroids alter cellular phenotype is the selective stimulation of transcription of specific genes [Yamamoto and Alberts, 1976]. My colleagues and I have pursued this aspect of glucocorticoid action in cultured mammalian cells, and we have focused on the expression of mammary tumor virus (MTV) DNA, a 9 kilobase (kb) pair element encoding a murine retrovirus. In general, viral gene expression is hormone-regulated in cells containing glucocorticoid receptors and exogenously introduced MTV sequences; in chronically infected rat HTC cells, for example, dexamethasone strongly stimulates production of MTV RNA [Ringold et al., 1977a,b].

The sole aspect of MTV RNA metabolism that is modulated by dexamethasone is the efficiency of transcription initiation. That is, the hormone-receptor complex stimulates MTV promoter utilization without altering the site of initiation, the sites or efficiencies of RNA splicing and polyadenylation [Ucker et al., 1983], or the rates of transcript elongation [Ucker and Yamamoto, 1984] and turnover [Ringold et al., 1978]. Moreover, regulation of MTV transcription by nonglucocorticoid hormones or other small molecules has not been observed. The relative simplicity of this control circuit in cultured cells has facilitated investigation of the molecular mechanism of glucocorticoid receptor action. It is also apparent from these and other studies that additional levels of control are often superimposed on the primary mode of steroid receptor action. For example, glucocorticoids regulate the expression of different genes in different target cells [Ballard et al., 1974], and the pattern of hormonal regulation is often modified drastically by other hormones and regulatory factors (see below).

This chapter summarizes briefly our current view of transcriptional regulation by glucocorticoids and describes related studies that might provide a general explanation for some apparently complex features of cell-specific and multihormonal control of gene expression. A model is presented that accounts for these and other findings; although largely speculative, the model is a useful pedagogical device for organizing and rationalizing information from diverse experimental approaches, and it suggests strategies for testing its facets.

II. RECEPTOR BINDING REGIONS ARE GLUCOCORTICOID RESPONSE ELEMENTS

Dexamethasone stimulates MTV transcription initiation half-maximally within 8–9 min of hormone addition [Ucker and Yamamoto, 1984], whereas bulk transcription is unaffected. The strong selectivity of this effect resides in the sequence-specific binding of the glucocorticoid-receptor complex to MTV DNA [Payvar et al., 1981, 1982, 1983; Govindan et al., 1982; Geisse et al., 1982; Pfahl, 1982; Scheidereit et al., 1983]. In vitro, purified receptor binds with high affinity to several discrete regions, one of them upstream of the MTV promoter and at least four located within transcribed sequences [Payvar et al., 1983]; in contrast, receptor fails to bind specifically to DNA whose expression is not glucocorticoid-regulated. The receptor associates with each region with a similar overall affinity, and each region appears to contain multiple receptor binding sites; for example, five discrete nuclease "footprints" reside between −305 bp and −84 bp relative to the MTV transcription initiation site [Payvar et al., 1983], defining "receptor binding region 1" (RB1; see Fig. 1A) within the MTV long terminal repeat (LTR).

Most importantly, DNA transfection experiments suggest that these receptor binding regions are likely to participate in mediating hormone-responsive transcription in vivo. Thus, transfections with various 5′ deletions of the LTR fused to "reporter" gene coding sequences revealed that removal of the RB1 region was accompanied by a reduction or loss of hormone responsiveness [Hynes et al., 1983; Groner et al., 1983; Majors and Varmus, 1983; Buetti and Diggelmann, 1983; Pfahl et al., 1983]. In addition, recombinant plasmids bearing receptor binding regions from transcribed portions of MTV DNA, but lacking known promoters, appeared somehow to confer hormone responsiveness on fortuitously positioned plasmid or cellular promoterlike elements [Payvar et al., 1982].

To test this notion directly, promoterless MTV DNA fragments were fused to the intact promoter and gene for herpes simplex virus thymidine kinase (tk), whose expression is not normally hormone-regulated (Fig. 1B). Thymidine kinase transcription was stimulated by glucocorticoids in chimeric constructions containing receptor binding sequences but not in fusions with MTV sequences that lacked binding activity [Chandler et al., 1983]. The sites of tk transcription initiation were not altered by the linked MTV fragments in either the presence or absence of hormone, demonstrating that the "glucocorticoid response element" (GRE) activities mapped by this procedure increase the efficiency of the heterologous promoter without affecting its specificity; similar results have now been obtained with fusions of MTV GRE-containing fragments to several different promoters [Majors and Varmus, 1983; Ringold, personal communication; Lee, personal communication].

As schematized in Figure 1B, fragments bearing RB1 are functional GREs when located in a variety of positions and in both orientations relative to the tk promoter; hormonal regulation was observed when RB1 was positioned as much as 19 bp closer and more than 1 kb farther upstream of the tk start site than it normally resides from the MTV start site [Chandler et al., 1983; DeFranco, Jones, and Tavtigian, unpublished results]. In addition, RB1 also functioned when placed more than 2 kb downstream of the tk promoter (DeFranco, unpublished results). Similarly, fragments bearing RB2 and RB3, or RB4, which normally reside 4.4–5.5 kb downstream from the MTV transcription start site, were competent GREs when fused upstream of the tk promoter (DeFranco and Jones, unpublished results). Thus the MTV GREs

A

B

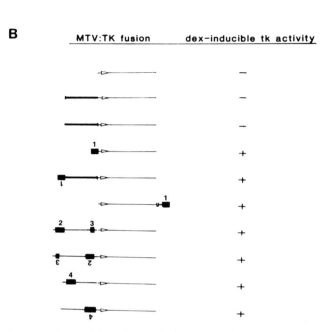

Fig. 1. Mapping and characterizing glucocorticoid response elements (GREs) in vivo. A, specific receptor binding regions identified in vitro within MTV DNA [Payvar et al., 1983]. Diagram shows 9 kb MTV proviral DNA in its normal integrated configuration; heavy lines denote LTR sequences; transcription initiates within the LTR at left and proceeds rightward (wavy line) to a polyadenylation site in the right LTR; numbered boxes depict specific binding regions RB1–RB5; square brackets demarcate a region not investigated in detail but known to contain at least two additional specific binding regions. B, assays of GRE activity using fusions of MTV DNA fragments to the intact tk promoter [Chandler et al., 1983; DeFranco, Jones and Tavtigian, unpublished results]. Various MTV DNA fragments were fused as shown to the intact tk promoter (open arrow) and the structural sequences for either tk or CAT. RB1 is normally centered 178 bp upstream from the MTV transcription initiation site; in these fusions, receptor binding regions were positioned 0.15–1.5 kb upstream of the tk start site and $\geqslant 2$ kb downstream. Rat XC cells were transfected with the recombinant plasmids, and promoter activity was assessed in stable or transient tranfectants in the presence or absence of 0.1 μM dexamethasone. The magnitudes of hormonal induction in these experiments were three- to 50-fold; however, quantitative comparisons of different stable transfectants are not meaningful; their extents of induction are influenced strongly by differences in copy number and chromosomal position [Feinstein et al., 1982; Chandler et al., 1983; Ucker et al., 1983]. Nevertheless, these results establish that (1) DNA fragments containing specific receptor binding regions in vitro carry GRE activity in vivo, (2) upstream and transcribed receptor binding regions are independently active as GREs, and (3) GREs are receptor-dependent transcriptional enhancers.

function in vivo as transcriptional enhancers (see below) whose activities appear to depend entirely on the presence of active receptor [Chandler et al., 1983; Yamamoto, 1984]; in the intact MTV element, we assume that the multiple GREs can operate additively to enhance initiation from the MTV promoter in the presence of dexamethasone.

To examine in detail the apparent correlation between receptor binding in vitro and hormone-dependent enhancement in vivo, an extensive series of defined mutations was constructed within RB1; for example, 21 mutants were isolated bearing clusters of from one to eight nucleotide substitutions (linker scanning mutations [McKnight and Kingsbury, 1982]) at discrete positions across the 222 bp RB1 region. Analyses of receptor footprinting activity in vitro and GRE activity in vivo [DeFranco et al., 1985; Wrange and DeFranco, unpublished results] revealed that the sequences essential for footprint formation are in every case also important for full GRE activity; moreover, the individual footprints appear to form independently in vitro, and each may contribute to the magnitude of hormone-dependent enhancement in vivo (Fig. 2). Moreover, sequences closely related to the consensus octanucleotide $AGA_T^ACAG_T^A$ appear to be essential (but perhaps not sufficient) for both activities [Payvar et al., 1983; DeFranco et al., 1985]. Finally, a 59 bp fragment that encompasses a single footprinted sequence from RB1 (see Fig. 2) retains both footprinting and GRE activities (Tavtigian, unpublished results). Taken together, these results establish genetically that the hormone dependence of GRE action reflects specific binding of the hormone-receptor complex to sequence elements that include the consensus octanucleotide.

III. STRUCTURAL CONSEQUENCES OF GRE ENHANCER ACTIVATION

Enhancers are DNA sequences within or near various genes transcribed by RNA polymerase II that are required for full activity of their cognate promoters [see Gluzman and Shenk, 1983]. Their distinguishing characteristics, discovered in the course of transfection experiments [Banerji et al., 1981; Moreau et al., 1981; Fromm and Berg, 1982], include the capacity to stimulate intact heterologous promoters to which they are fused, to function more efficiently within cells from particular species and/or cell types, to operate preferentially on the most proximal promoter, and to act in either orientation and over a broad range of distance relative to the affected promoter. To account for these properties, it was proposed that enhancers might comprise regions of special DNA configuration and/or chromatin structure that specifically favor promoter function, perhaps by providing "bidirectional

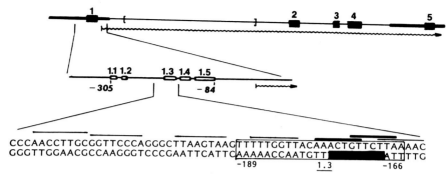

CCCAACCTTGCGGTTCCCAGGGCTTAAGTAAC TTTTTGGTTACAAACTGTTCTTAA AAC
GGGTTGGAACGCCAAGGGTCCCGAATTCATTC AAAAACCAATGTT ▮▮▮▮ ATTTTG
                                 -189   1.3   -166

Fig. 2. Identification of DNA sequences involved in specific receptor binding and GRE enhancer action [Payvar et al., 1983; DeFranco et al., 1984; Wrange, DeFranco, Jones, and Tavtigian, unpublished results]. Upper diagram depicts MTV proviral DNA as described in legend to Figure 1. Middle diagram shows a region of the LTR that includes RB1, which is defined by five nuclease footprints 1.1–1.5 (open boxes) between −84 bp and −305 bp from the MTV transcription start site; lower diagram shows the DNA sequence of a 59 bp fragment encompassing footprint 1.3 (boxed). The consensus octanucleotide (see text) is shaded. Overlinings represent individual linker scanning mutations in which the specified wild type nucleotides have been substituted by a synthetic SacI octanucleotide; fine overlinings correspond to linker scanning mutations that lack detectable phenotypes; heavy overlinings depict mutations with phenotypic effects. Footprinting and transfections with wild type and mutant RB1 yielded the following general conclusions: 1) Mutations outside of footprinted sequences have no effect. 2) *Some* mutations within footprinted sequences alter receptor binding and enhancement. 3) All mutations that reduce receptor binding also reduce enhancement. 4) Each footprint is formed independently, and each appears to contribute to enhancement. 5) A single footprint sequence is necessary and sufficient for footprint formation and enhancement. 6) Only mutations that alter the consensus octanucleotide $AGA_T^ACAG_T^A$ affect footprint formation and enhancement.

entry sites" for RNA polymerase or other transcription factors [Benoist and Chambon, 1981; Wasylyk et al., 1983]. Consistent with this notion, discrete regions of chromatin that are hypersensitive to DNaseI and other nucleases have been observed at or near a variety of enhancers; in particular, such structural features were detected near the SV40, polyoma, and immunoglobulin enhancers only within cell types in which promoters under their control are active or potentially active [Cremisi, 1981; Saragosti et al., 1982; Herbomel et al., 1981; Parslow and Granner, 1982; Weischet et al., 1982; Mills et al., 1983; Chung et al., 1983].

DNaseI sensitivity studies of integrated glucocorticoid-responsive MTV-tk fusion genes [Zaret and Yamamoto, 1984] revealed two classes of chromatin structure alterations that accompany hormone treatment (Fig. 3). First, discrete hypersensitive regions appeared at sites of specific receptor:DNA

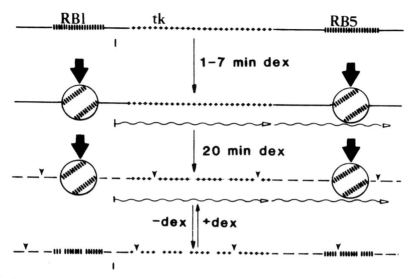

Fig. 3. Reversible and persistent changes in chromatin structure accompany enhancer activation [Zaret and Yamamoto, 1984]. Diagrams show the structure of an MTV derivative in which most of the body of the provirus has been deleted and replaced with the tk coding sequence (stippled line) flanked by the MTV LTRs, which contain RB1 and RB5 (striped lines) as well as the intact MTV promoter. Transcription starts in the LTR at left and proceeds rightward (wavy lines) to either the tk or the right LTR polyadenylation signal. Stably transfected cell lines bearing one or more integrated copies of this construction were exposed to dexamethasone; after various times, nuclei were isolated and treated briefly with DNaseI. Two regions of DNaseI hypersensitivity (large arrows) were detected after a few minutes of hormone treatment and mapped at RB1 and RB5, where receptor (large circles) binds specifically in vitro. Subsequently, increased moderate DNase sensitivity was detected within the transcribed region as well as upstream (small arrows denote overall nuclease sensitivity rather than specific cleavage loci). The hypersensitive regions disappeared upon hormonal withdrawal when tk expression was shut off, but the moderate DNase sensitivity persisted.

interactions; the hypersensitive structures were established within a few minutes after hormone addition, apparently preceding the increased rate of transcription initiation at the MTV promoter. These structures disappeared upon hormone removal. Second, the moderate DNaseI sensitivity of the entire chromatin domain encompassing the gene increased and remained elevated after hormone withdrawal, when tk expression was shut off. The potential significance of the latter structural transition has not been investigated; a similar persistent structural alteration at the ovalbumin locus after initial exposure to estrogen [Shepard et al., 1980; Lawson et al., 1982] might, in principle, correlate with distinct kinetics and hormone specificity of the secondary estrogen response [Hagar et al., 1980]. In any case, parallel studies with the GRE mutants described above should help to define in detail the relationship between receptor binding in vitro, establishment of induced hypersensitive regions in nuclei, and GRE-mediated enhancement in vivo. Taken at face value, the present results imply that the specific receptor:DNA interaction can directly create an active transcriptional enhancer by specifically altering the local chromatin structure.

IV. ENHANCERS AND REGULATORY EVOLUTION

It seems likely that response elements for other steroids also function as hormone-dependent enhancers. Indeed, the normal function of all identified enhancer elements might be to bind and mediate the action of a class of specific transcriptional regulatory proteins functionally analogous to the glucocorticoid receptor [Yamamoto et al., 1983; Yamamoto, 1984]; numerous additional sequence elements that effect positive regulation on eukaryotic promoters appear to be controlled by diffusible factors and to share some of the key properties of enhancers [Errede et al., 1981; Williamson et al., 1981; Hinnebusch and Fink, 1983; Guarente et al., 1984]. Thus enhancement might be a common or predominant mode of transcriptional regulation in eukaryotes, conceivably reflecting a fundamental aspect of the evolution of complex organisms [Yamamoto, 1983].

In a theoretical discussion of the evolution of gene networks regulated by conserved signalling molecules, Tomkins [1975] suggested that a given symbol must somehow acquire new functions appropriate to the particular needs of each organism but that, once it interconnected several important cellular processes, the overall consequences of its specific actions would be evolutionarily constrained, as with the universality of the genetic code. How might complex regulatory networks maintain stably their general contexts while simultaneously arising and evolving at rates consistent with the morphological and physiological diversity of different organisms? The striking properties of GRE enhancers suggest a simple way to accommodate these requirements: low-frequency transpositions in germ line DNA could stochastically position a GRE sufficiently close to a given promoter to confer upon it the potential for hormonal regulation. The important points here are that promoters and regulatory elements need not coevolve and that the gene rearrangement need not bring the two elements into a rigorously prescribed spatial or orientational relationship. As with all alterations in germ line DNA, unfavorable transposition events would be removed through selection, neutral transpositions would coexist within the wild type population, and rearrangements that produce favorable additions to a regulated gene network would be fixed, perhaps eventually undergoing secondary rearrangements that more closely juxtapose the GRE and promoter.

Wilson and coworkers [1977] have proposed that phenotypic evolution is driven by shifts in gene regulation rather than by structural sequence alterations. This notion has been documented in bacterial populations and can be inferred in higher organisms from species-specific differences in the quantitative, temporal, or spatial expression of particular gene products and from differences in responses to steroids and other regulatory signals. Most importantly, rates of phenotypic evolution appear to correlate with rates of gene rearrangement. I suggest that the crucial rearrangements involve not only the duplication of structural sequences but also the transposition of regulatory elements such as GREs.

The multiple GRE elements within MTV DNA could represent multiple rearrangement events that together yield the strong hormone responsiveness observed; preliminary studies imply that other steroid regulated genes also contain multiple specific receptor binding regions [Mulvihill et al., 1982; Bailly et al., 1983]. In a like manner, two or more enhancers regulated by *distinct* activator proteins might be transposed into the vicinity of a single promoter. Such "stacking" of functional response elements is readily accommodated by the properties of enhancers and, as discussed below, could provide an efficient means for modulating gene expression with multiple effectors.

V. HORMONE-REGULATED GENE NETWORKS

Operationally, the set of genes that is *potentially* responsive to glucocorticoids is subdivided during development into cell-specific gene networks. Thus expression of a particular gene might be hormone-regulated in one cell and completely suppressed in another. For example, rat growth hormone gene transcription is induced in pituitary cells by glucocorticoids [Evans et al., 1982], whereas no growth hormone RNA is produced in liver cells. Indeed, the magnitude of inactivation is extraordinary: Ivarie et al. [1983] showed that growth hormone is at least eight orders of magnitude less abundant in liver than in pituitary. This repression process, some 10^5-fold stronger than that effected by bacterial repressor proteins, probably reflects establishment and maintenance of chromatin conformations that preclude both basal and hormone-stimulated promoter activity. A complete blockade of expression by chromatin structure cannot, however, explain other features of cell-specific gene regulation by steroids. For example, transcription of the chicken transferrin gene is stimulated by estradiol in oviduct, but it is constitutively expressed in liver [Lee et al., 1978] despite the presence of competent estradiol receptors in both tissues.

Whatever its molecular basis, determination of cell-specific gene networks can be viewed as a *static control* that is superimposed stably on the activity of a given hormone response element. In addition, there are *dynamic controls* that can exert profound modulatory effects. Thus, hormone-regulated genes are commonly controlled by multiple hormones and factors, and the characteristics of multifactor regulation are often more complex than the summed responses to each regulator given separately. For example, synergistic and permissive effects of two or more hormones on the expression of single gene are common, as is illustrated by the finding that estradiol induction of ovalbumin transcription in oviduct tissue occurs only in the presence of insulin, whereas neither hormone alone has a marked effect [Evans et al., 1981; McKnight, personal communication]. In contrast, oviduct transferrin expression is induced by estradiol alone, demonstrating that estrogen receptor function per se is not insulin-dependent. Similarly, in rat HTC cells, cAMP alone fails to stimulate expression of tyrosine aminotransferase, but dexamethasone permits cAMP to act as a strong inducer [Granner, 1976]. Conversely, glucocorticoids have no effect on plasminogen activator mRNA production in HTC cells, whereas prior or simultaneous treatment with cAMP permits strong dexamethasone induction of plasminogen activator RNA synthesis [Gelehrter et al., 1983].

Little is known about the nature of permissive and synergistic effects of any two (or more) hormones. Posttranscriptional and/or posttranslational phenomena [e.g., see Firestone et al., 1982] could contribute in some cases to overall patterns of multifactor regulation, but it is clear that these complex phenomena can occur solely at the level of transcription.

VI. COMBINATORIAL EFFECTS OF ENHANCERS

Exploiting the properties of the GRE as a transcriptional enhancer, we have begun to examine the possibility that a pair of functionally distinct enhancers might mimic aspects of both static and dynamic controls [DeFranco et al., 1984; DeFranco, unpublished results]. For example, Figure 4 summarizes results obtained with plasmids bearing a GRE and/or the rat chymotrypsin enhancer (CHY) [see Walker et al., 1983; Boulet and Rutter, unpublished results], both in *cis* to the MTV promoter (P_{MTV}) or the thymidine kinase promoter (P_{TK}), which drive expression of the bacterial chloramphenicol acetyltransferase (CAT) coding sequences. These plasmids were

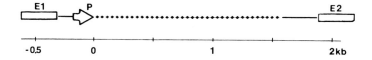

Plasmid	Enhancer$_1$	Promoter	Enhancer$_2$	Relative CAT Activity			
				XC cells		AR4-2J cells	
				−dex	+dex	−dex	+dex
OTCY	−	tk	CHY	1	1	10	10
GTCO	GRE	tk	−	1	3	1	2
GTCY	GRE	tk	CHY	1	3	10	80
GMCO	GRE	MTV	−	1	4	1	50
GMCY	GRE	MTV	CHY	1	4	1	280

Fig. 4. Combinatorial actions of pairs of enhancers: effects of promoter and cell type [DeFranco et al., 1984; DeFranco, unpublished results]. A simple plasmid was constructed that allows insertion of a promoter-bearing fragment upstream of the bacterial chloramphenicol acetyltransferase gene as well as integration of enhancer-bearing fragments both upstream and downstream of the gene. Magnitude of promoter function was inferred from CAT activity monitored in extracts of transiently transfected cells and from direct assays of properly initiated RNA by primer extension. Results are shown for transfections with plasmids containing GRE and/or CHY (see text) in cis with promoter fragments from either TK (P_{TK}) or MTV (P_{MTV}). Transfections were carried out in the presence or absence of dexamethasone (GRE active or inactive, respectively), in rat exocrine pancreas AR4-2J cells or rat fibroblast XC cells (CHY active or inactive, respectively). These data can be readily interpreted in terms of the model presented in Figure 5 (see below). For example: 1) with P_{TK}, the expected cell type specificity of CHY is observed (GTCO and GTCY in AR4-2J). Surprisingly, CHY is cryptic when P_{MTV} is substituted (GMCO and GMCY in AR4-2J), as if different factors limit the initiation rate of the two promoters. 2) With p_{TK}, synergy is observed when both GRE and CHY are active, as if the two enhancers provide different rate-limiting factors that together might form a functional complex. 3) With P_{MTV}, GRE is far more effective in AR4-2J cells than in XC, as if the latter cell line contains a more active pool of a GRE-specific transcription factor. 4) With P_{MTV} in AR4-2J cells, CHY activity is detectable only after GRE activation, as if GRE provides the primary rate-limiting factor for MTV promoter function and CHY relieves a secondary limitation.

transfected into rat fibroblast XC cells or into AR4-2J cells, an established line derived from rat exocrine pancreas. CHY activity has been observed only in exocrine pancreas cells [Walker et al., 1983], whereas GRE is potentially active in both lines after dexamethasone treatment. As is shown in Figure 4, the paired enhancers stimulated promoter activity synergistically in certain combinations (e.g., compare plasmids GTCO and GTCY in AR4-2J cells), whereas in others GRE action was permissive for function of the otherwise cryptic CHY enhancer (compare GMCO and GMCY in AR4-2J cells). Clearly, the pattern of CAT expression and hormonal response changed both quantitatively and qualitatively when the promoter or cell type was changed; indeed, the patterns appear reminiscent of those observed with genes under multifactor control.

Given the results with these and other simple constructions (DeFranco, unpublished results), it would be interesting to test with a normal cellular gene the notion that pairs of heterologous enhancers give rise to complex regulatory behavior. Several genes that are regulated by multiple hormones have been cloned and expressed in transfected cells [Kurtz, 1981; Robins et al., 1982; Renkawitz et al., 1982; Lai et al., 1983], although the sites and activities of response elements for two or more transcriptional regulators associated with such a candidate gene have not yet been determined in detail. In principle, however, it should be possible to assess receptor occupancy and chromatin configuration of the response element for one hormone in the presence and absence of the other. Eventually, of course, a full understanding of the functional relationships between regulators will require cell-free reconstruction of the regulatory events.

Taken at face value, however, the present results can be interpreted in terms of a model in which different enhancer elements, each activated by specific diffusible regulators (Fig. 5A), potentiate the action of *different* transcription factors (Fig. 5B); this new assumption accounts for the different efficacies of an enhancer on different promoters, since the sequence of each promoter determines which factors intrinsically limit its efficiency. Such a scheme, as illustrated in Figure 5C, readily gives rise to synergism, permissive effects, and cell-specific constitutivity vs. regulation. Thus this model suggests a simple and specific mechanism by which a limited number of regulatory components might yield complex networks of control. As such, it represents a refinement of an earlier hypothesis for steroid receptor action [Yamamoto and Alberts, 1976] and draws conceptually on general schemes for combinatorial gene regulation envisioned by Britten and Davidson [1971], Georgiev [1969], Gierer [1973], and others.

VII. CONCLUSIONS AND PERSPECTIVES

Indirect experiments imply that many or all enhancer elements, like GREs, are activated by specific trans-acting factors analogous to the glucocorticoid receptor; identification and functional analysis of other enhancer binding proteins will likely confirm that enhancers are not simply elements essential for optimal constitutive expression but rather mediate an important general mode of transcriptional regulation. Conceivably, the apparent widespread utilization of this mechanism could reflect the capacity of enhancers to function from a variety of positions, thereby allowing efficient evolution of regulated gene networks and a simple solution to the problem of multifactor gene regulation.

The defining characteristics of enhancers and enhancement remain largely phenomenological; undoubtedly, common and unique features of different enhancers will soon be distinguished at the molecular level. Clearly, the investigation of glucocorticoid receptor-regulated transcription is a useful approach to elucidating the nature of one such element. The receptor is perhaps the most fully characterized eukaryotic positive transcriptional regulatory factor, and it mediates the action of a defined, physiologically important ligand. As expected, it is present in low relative abundance (less than .01% of total cell protein), but it has been purified to near-homogeneity [Wrange et al., 1979, 1984], polyclonal and monoclonal antibodies have been prepared [Okret et al., 1981, 1984; Eisen et al., 1981; Westphal et al., 1982], and cDNAs carrying receptor coding sequences have recently been isolated [Miesfeld et al., 1984]. Moreover, hundreds of somatic cell mutants bearing a variety of phenotypically distinct defects in receptor function have been isolated and characterized [for review, see Stevens et al., 1983]. Finally, the specific binding sites for receptor have been defined at the DNA sequence level both biochemically and genetically, and they correspond to the sequences involved in GRE enhancer activity [DeFranco et al., 1985].

It is curious that GREs and other enhancers appear to encompass rather substantial segments of DNA, often on the order of 200 bp or more, and both substitution and deletion mutational studies imply that subregions across the entire element seem to contribute independently to the overall enhancer activity. In the case of the GRE at the RB1 region of MTV, just one of the five sequences footprinted by receptor in that region is sufficient to yield GRE activity. Why are GREs and other enhancers organized in an apparently degenerate manner? Conceivably, even subtle differences in the efficiency of formation, stability, or precise configuration of the altered chromatin structures corresponding to enhancer activation could modify responses in ways that are difficult or impossible to detect in current assays. Defining the

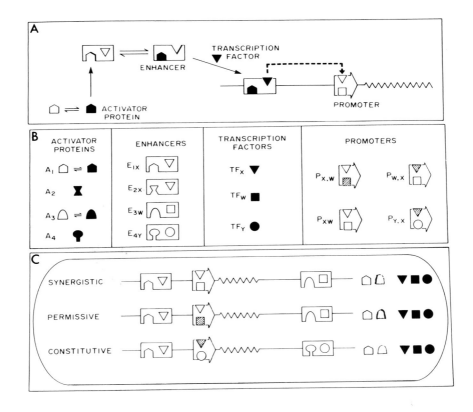

precise biochemical nature of the structural alterations seems likely to be crucial to understanding the mechanism of enhancement itself.

Simple plasmid constructions bearing pairs of heterologous enhancers appear to recapitulate aspects of complex physiological regulation by two or more hormones or other factors. This could indicate that a given active enhancer structure is recognized by a specific factor, or group of factors, that limits the initiation rates of a particular class of promoters; alternatively, the structural change might be propagated along the chromosome, modifying promoter structures such that certain of them would interact more productively with particular transcription factors. Such factors could include RNA polymerase itself, proteins that associate transiently with RNA polymerase, or species that function in initiation without directly contacting the polymerization machinery. The important point is that the apparent efficiency of a given enhancer will depend both on the cell type in which the enhancer is assayed and on the promoter to which it is juxtaposed. Thus the relative levels of relevant activator proteins and transcription factors, and the overall chromatin structure of the enhancer and promoter region, would be cell type-

---

Fig. 5. A model for enhancer action and multifactor gene regulation. A, the GRE enhancer is active only in the presence of specifically bound glucocorticoid receptor, which itself functions only in the presence of bound hormone. Glucocorticoid receptor is the prototypical enhancer activator protein, and indirect evidence suggests that the activities of many or all enhancers are mediated by trans-acting factors. Receptor binding results in the formation of an altered chromatin structure at the GRE that is presumed to potentiate the activity of a transcription factor. One possibility is that the transcription factor binds at the active enhancer, subsequently translocating in either direction along the chromosome until it reaches a promoter. B, the assumption crucial to the model is that different activator protein:enhancer complexes might be specifically recognized by *different* specific transcription factors. Promoters are sequence elements that have the intrinsic capacity to "capture" the factors and enzymes necessary to initiate RNA synthesis; their different primary sequences determine which transcription factors are captured least efficiently and are therefore rate limiting for initiation efficiency. Open symbols within each promoter represent transcription factors that limit initiation rate, whereas hatched symbols denote the factor that is the next least efficiently captured. That is, $P_{X,W}$ is rate limited by TF_X; if that limitation were relieved, $P_{X,W}$ would then be limited by TF_W; in contrast, P_{XW} is rate limited in its capacity to capture both TF_X and TF_W. C, a hypothetical cell is shown that contains all three transcription factors, two of the four activator proteins, and three genes, each with unique promoter and enhancer combinations, but all containing E_{1X}. Note that the pattern of expression of the three genes differs dramatically when the receptorlike activator proteins are activated. The two enhancers stimulate the *upper gene* independently and synergistically; E_{3W} action on the *middle gene* requires a permissive effect mediated by E_{1X}. Finally, the *lower gene* is constitutively expressed; E_{4Y} is without effect owing to the cell-specific absence of A_4, and E_{1X} provides a factor that is not rate limiting under these conditions. Note that the lower gene would be permissively regulated in a cell type containing A_4 and A_1.

specific, and the potential activity of the enhancer would be inversely related to the intrinsic capacity of the promoter to "capture" that factor in the absence of the enhancer. The model makes some rather clear predictions that could be indirectly tested in vivo in "commitment" or competition assays [Bogenhagen et al., 1982; Scholer and Gruss, 1984; Brown, 1984] and should eventually be directly testable in cell-free extracts in vitro. Together, approaches such as those outlined here might begin to provide insight into the mechanisms by which a restricted number of regulatory molecules generates and coordinates the diverse programs of gene expression that govern development and physiological function.

ACKNOWLEDGMENTS

I am grateful to the past and present colleagues in my laboratory whose skill and ingenuity have produced the data and seeded the ideas that are summarized here. In addition to acknowledging those who contributed to published studies cited below, I thank Don DeFranco, Susan Jones, Sean Tavtigian, Orjan Wrange, and Ken Zaret for permission to mention some of their unpublished results. I also acknowledge Don DeFranco, Sandro Rusconi, and Ken Zaret for their critiques of the manuscript and Kathleen Rañeses for its expert preparation. Our research is supported by grants from the National Institutes of Health and the National Science Foundation; K.R.Y. is recipient of a Teacher-Scholar award from the Henry and Camille Dreyfus Foundation.

VIII. REFERENCES

Anderson JN (1984): In Goldberger RF, Yamamoto KR (eds): "Biological Regulation and Development. Vol 3B. Hormone Action." New York: Plenum Press, pp 169–212.

Bailly A, Atger M, Atger P, Cerbon M-A, Alizon M, Hai MTV, Logeat F, Milgrom E (1983): J Biol Chem 258:10384–10389.

Ballard PL, Baxter JD, Higgins SJ, Rousseau GG, Tomkins GM (1974): Endocrinology 94:998–1002.

Banerji J, Rusconi S, Schaffner W (1981): Cell 27:299–308.

Benoist C, Chambon P (1981): Nature 290:304–310.

Bogenhagen DF, Wormington WM, Brown DD (1982): Cell 28:413–421.

Britten RJ, Davidson EH (1971): Q Rev Biol 46:111–138.

Brown DD (1984): Cell 37:359–365.

Buetti E, Diggelmann H (1983): EMBO J 2:1423–1429.

Burshell A, Stathis PA, Do Y, Miller SC, Feldman D (1984): J Biol Chem 259:3450–3456.

Chandler VL, Maler BA, Yamamoto KR (1983): Cell 33:489–499.

Chung S, Folsom V, Wooley J (1983): Proc Natl Acad Sci USA 80:2427–2431.

Cremisi C (1981): Nucleic Acids Res 9:5949–5964.

DeFranco D, Wrange O, Merryweather J, Yamamoto KR (1985): In Herskowitz I, Simon M (eds): "Genome Rearrangement" (UCLA Symposium on Molecular and Cellular Biology, New Series). New York: Alan R. Liss, Inc., pp. 305–321.

Eisen HJ, Schleenbaker RE, Simons SS (1981): J Biol Chem 256:12920–12925.

Errede B, Cardillo TS, Wever G, Sherman F (1981): Cold Spring Harbor Symp Quant Biol 45:593–602.

Evans MI, Hager LJ, McKnight GS (1981): Cell 25:187–193.

Evans RM, Birnberg NC, Rosenfeld MG (1982): Proc Natl Acad Sci USA 79:7659–7663.

Feinstein SC, Ross SR, Yamamoto KR (1982): J Mol Biol 156:549–566.

Firestone GL, Payvar F, Yamamoto KR (1982): Nature 300:221–225.

Fromm M, Berg P (1982): J Mol Appl Genet 1:457–481.

Geisse S, Scheidereit C, Westphal HM, Hynes NE, Groner B, Beato M (1982): EMBO J 1:1613–1619.

Gelehrter TD, Barouski-Miller PA, Coleman PL, Cwikel BJ (1983): Mol Cell Biochem 53/53:11–21.

Georgiev GP (1969): J Theor Biol 25:473–490.

Gierer A (1973): Cold Spring Harbor Symp Quant Biol 38:951–961.

Gluzman Y, Shenk T (1983): "Enhancers and Eukaryotic Gene Expression." Cold Spring Harbor, New York: Cold Spring Harbor Laboratory.

Govindan MV, Spiess E, Majors J (1982): Proc Natl Acad Sci USA 79:5157–5161.

Granner DK (1976): Nature 259:572–573.

Groner B, Hynes NE, Rahmsdorf U, Ponta H (1983): Nucleic Acids Res 11:4713–4725.

Guarente L, Lalonde B, Gifford R, Alani E (1984): Cell 36:503–511.

Hager LJ, McKnight GS, Palmiter RD (1980): J Biol Chem 255:7296–7800.

Herbomel P, Saragosti S, Blangy D, Yaniv M (1981): Cell 25:651–658.

Hinnebusch AG, Fink GR (1983): J Biol Chem 258:5238–5247.

Hynes N, van Ooyen AJJ, Kennedy N, Herrlich P, Ponta H, Groner B (1983): Proc Natl Acad Sci USA 80:3637–3641.

Ivarie RD, Schachter BS, O'Farrell PH (1983): Mol Cell Biol 3:1460–1467.

Kurtz DT (1981): Nature 291:629–631.

Lai EC, Riser ME, O'Malley BW (1983): J Biol Chem 258:12693–12701.

Lawson GM, Knoll BJ, March CJ, Woo SL, Tsai M-J, O'Malley BW (1982): J Biol Chem 257:1501–1507.

Lee DC, McKnight GS, Palmiter RD (1978): J Biol Chem 253:3494–3503.

Loose DS, Schurman DJ, Feldman D (1981): Nature 293:477–479.

Majors J, Varmus HE (1983): Proc Natl Acad Sci USA 80:5866–5870.

McKnight SL, Kingsbury R (1982): Science 217:316–324.

Miesfeld R, Okret S, Wikstrom A-C, Wrange O, Gustafsson J-A, Yamamoto KR (1984): Nature 312:779–781.

Mills FC, Fisher LM, Kuroda R, Ford AM, Gould HJ (1983): Nature 306:809–812.

Moreau P, Hen R, Wasylyk B, Everett R, Gaub MP, Chambon P (1981): Nucleic Acids Res 9:6047–6067.

Mulvihill ER, LePennec J-P, Chambon P (1982): Cell 28:621–642.

Okret S, Carlstedt-Duke J, Wrange O, Carlstrom K, Gustafsson J-A (1981): Biochim Biophys Acta 677:205–219.

Okret S, Wikstrom A-C, Wrange O, Andersson B, Gustafsson J-A (1984): Proc Natl Acad Sci USA 81:1609–1613.

Parslow TG, Granner DK (1982): Nature 299:449–451.

Payvar F, DeFranco D, Firestone GL, Edgar B, Wrange O, Okret S, Gustafsson J-A, Yamamoto KR (1983): Cell 35:381–392.

Payvar F, Firestone GL, Ross SR, Chandler VL, Wrange O, Carlstedt-Duke J, Gustafsson J-A, Yamamoto KR (1982): J Cell Biochem 19:241–247.

Payvar F, Wrange O, Carlstedt-Duke J, Okret S, Gustafsson J-A, Yamamoto KR (1981): Proc Natl Acad Sci USA 78:6628–6632.

Pfahl M (1982): Cell 31:475–482.

Pfahl M, McGinnis D, Hendricks M, Groner B, Hynes NE (1983): Science 222:1341–1343.

Renkawitz R, Beug H, Graf T, Matthias P, Grez M, Schutz G (1982): Cell 31:167–176.

Ringold GM, Cardiff RD, Varmus HE, Yamamoto KR (1977a): Cell 10:11–18.

Ringold GM, Shank PR, Yamamoto KR (1978): J Virol 26:93–101.

Ringold GM, Yamamoto KR, Bishop JM, Varmus HE (1977b): Proc Natl Acad Sci USA 74:2879–2883.

Robins DM, Paek I, Seeburg PH, Axel R (1982): Cell 29:623–631.

Saragosti S, Cereghini S, Yaniv M (1982): J Mol Biol 160:133–146.

Scheidereit C, Geisse S, Westphal HM, Beato M (1983): Nature 304:749–752.

Scholer HR, Gruss P (1984): Cell 36:403–411.

Shepherd JH, Mulvihill ER, Thomas PS, Palmiter RD (1980): J Cell Biol 87:142–151.

Stevens J, Stevens Y-W, Haubenstock H (1983): In Litwack G (ed): "Biochemical Actions of Hormones. Vol 10." New York: Academic Press, pp 383–446.

Timberlake WE, Orr, WC (1984): In Goldberger RF, Yamamoto KR (eds): "Biological Regulation and Development. Vol 3B. Hormone Action." New York: Plenum Press, pp 255–283.

Tomkins GM (1975): Science 189:760–763.

Ucker DS, Firestone GL, Yamamoto KR (1983): Mol Cell Biol 3:551–561.

Ucker DS, Yamamoto KR (1984): J Biol Chem 259:7416–7420.

Walker MD, Edlund T, Boulet AM, Rutter WJ (1983): Nature 306:557–561.

Wasylyk B, Wasylyk C, Augereau P, Chambon P (1983): Cell 32:503–514.

Weischet WO, Glotov BO, Schnell H, Zachau HG (1982): Nucleic Acids Res 10:3627–3645.

Westphal HM, Moldenhauer G, Beato M (1982): EMBO J 1:1467–1471.

Williamson VM, Young ET, Ciriacy M (1981): Cell 23:605–614.

Wilson AC, Carlson SS, White TJ (1977): Annu Rev Biochem 46:573–639.

Wrange O, Carlstedt-Duke J, Gustafsson J-A (1979): J Biol Chem 254:9284–9290.

Wrange O, Okret S, Radojcic M, Carlstedt-Duke J, Gustafsson J-A (1984): J Biol Chem 259:4534–4541.

Yamamoto KR (1983): In Erikkson H, Gustafsson J-A (eds): "Steroid Hormone Receptors: Structure and Function" (Nobel Symposium 57). Amsterdam: Elsevier/North Holland Biomedical Press, pp 285–304.

Yamamoto KR (1984): In Ginsberg HS, Vogel HJ (eds): "Transfer and Expression of Eukaryotic Genes." New York: Academic Press, pp 79–92.

Yamamoto KR, Alberts BM (1976): Annu Rev Biochem 45:721–746.

Yamamoto KR, Maler BA, Chandler VL (1983): In Gluzman Y, Shenk T (eds): "Enhancers and Eukaryotic Gene Expression." Cold Spring Harbor, New York: Cold Spring Harbor Laboratory, pp 165–169.

Zaret KS, Yamamoto KR (1984): Cell 38:29–38.

Molecular Developmental Biology, pages 149–159

Developmental Changes in the Response of Mouse Eggs to Injected Genes

Howard Y. Chen, Myrna E. Trumbauer, Karl M. Ebert, Richard D. Palmiter, and Ralph L. Brinster

Laboratory of Reproductive Physiology, School of Veterinary Medicine, University of Pennsylvania, Philadelphia, Pennsylvania 19104 (H.Y.C., M.E.T., K.M.E., R.L.B.), and Department of Biochemistry, Howard Hughes Medical Institute, University of Washington, Seattle, Washington 98195 (R.D.P.)

I. INTRODUCTION

During the past several years, we have examined the capability of mouse eggs of early developmental stages to translate foreign messenger RNA microinjected into the cytoplasm [Brinster et al., 1980, 1981b] and to transcribe new genes injected into the nucleus [Brinster et al., 1981a, 1982]. These studies have provided information about the capability of early mammalian egg cells to process RNA and DNA molecules. The studies on DNA have shown that two RNA polymerase type III genes, the *Xenopus* 5S ribosomal RNA gene and the adenovirus-associated VA gene, are both transcribed much more effectively in the growing oocyte and mature oocyte than in the fertilized ovum [Brinster et al., 1981a]. In addition, it has been demonstrated that two RNA polymerase type II genes, the herpes simplex virus (HSV) thymidine kinase (TK) gene and a hybrid MK gene in which the HSV TK structural gene was fused to the mouse metallothionein-I (MT) promoter, were transcribed and expressed in fertilized mouse eggs [Brinster et al., 1982]. However, in contrast to genes transcribed by polymerase III, we found that the fertilized egg has a much greater ability than the oocyte to express injected genes transcribed by polymerase II.

These findings and related results on the processing and stability of injected messenger RNA indicate that there are dramatic changes in the ability of the mouse egg to produce and process messenger RNA immediately following fertilization. These profound changes in nucleic acid metabolism are in sharp contrast to the stability of other metabolic parameters, such as total RNA synthesis (as assayed by tritiated uridine incorporation) protein turnover, and oxygen consumption [Brinster, 1973; Sherman, 1979; Bachvarova, 1985].

II. EFFECT OF DEVELOPMENTAL STAGE ON INJECTED GENE TRANSCRIPTION

Because mouse oocytes transcribe RNA polymerase III genes more effectively than fertilized eggs, we compared the ability of several early developmental egg stages to transcribe polymerase type II genes. As a test gene for these experiments we chose the HSV TK gene. Approximately 60,000 copies were injected into the nucleus of each cell. This number is slightly higher than that previously shown to result in maximal TK activity in the fertilized one-cell ovum [Brinster et al., 1982]. The slight excess was chosen to ensure that adequate gene copies were available in the event that any of the other stages examined produced a higher level of activity than fertilized ova.

Table I shows the results of assays performed on growing oocytes, mature oocytes, unfertilized ova, and fertilized ova. It is clear from the results that the growing oocyte, mature oocyte, and fertilized ovum are all capable of transcribing the injected gene and that the messenger RNA is translated into active enzyme. The unfertilized ovum is arrested in the second meiotic metaphase and does not contain a nucleus; therefore, the gene was injected into the cytoplasm. Very little TK enzyme was found. Cytoplasmic injection of the gene in the fertilized egg also results in little or no TK activity even though nuclear injection is associated with very high levels of TK activity. The most dramatic finding is the enormous increase (30-fold) in the TK levels found in the fertilized eggs compared to mature oocytes, a developmental period of only 12–24 hr.

The higher TK activity in fertilized eggs could result from increased 1) DNA stability, 2) transcription, 3) mRNA stability, 4) enzyme synthesis, or 5) enzyme stability. Therefore, we examined these factors as possible causes. Figure 1 shows a comparison of the DNA present in growing oocytes and fertilized one-cell ova at several times following injection of approximately equal amounts of DNA into the two stages. Although there might be small variations in the form of the DNA between the stages, it is clear that there is

TABLE I. Herpes Simplex Virus Thymidine Kinase Activity in Early Stages of Mouse Ova Injected With the Cloned Gene for the Enzyme

Stage of development[a]	Site of injection[b]	Plasmid injected[c]	Formation of [³H]TMP (cpm × 10⁻³)
Growing oocytes	—	None	2.29 ± 0.31[d]
	GV	pBR322	2.66 ± 0.28
	GV	pTK	34.02 ± 2.46
Mature oocytes	—	None	6.14 ± 0.56
	GV	pTK	55.11 ± 17.59
Unfertilized ova	—	None	6.07 ± 0.79
	C	pTK	9.62 ± 2.75
Fertilized	—	None	8.73 ± 1.21
	PN	pBR322	6.72 ± 1.00
	PN	pTK	$1,580 \pm 150$
	C	pTK	8.40 ± 0.67

[a]Growing oocytes were dissected from the ovaries of 14-day-old females and were 50–60 μm in diameter. Mature oocytes (just prior to ovulation) were dissected from ovaries of 8-week-old females. Unfertilized and fertilized ova were flushed from the oviducts at 9–12 hr after ovulation [See Brinster et al., 1981[b]].

[b]GV, germinal vesicle; C, cytoplasm; PN, pronucleus.

[c]The plasmids injected were pBR322, the vector plasmid with no cloned gene; pTK, which contains the 3.5 kb herpes simplex viral thymidine kinase gene Bam fragment in pBR322; 60,000 plasmids were injected per nucleus.

[d]Following injection, the ova were cultured in [³H]thymidine for 22 hr, and [³H]thymidine monophosphate ([³H]TMP), the reaction product of thymidine kinase, was then assayed [Brinster et al., 1982]. Values are for 25 ova and are mean $\pm$ SEM for three or more experiments.

not a dramatic difference in stability of the injected DNA. Much of the DNA is still present in its injected form after the eggs have been cultured for 24 hr. There appears to be a slight conversion from relaxed circles to supercoiled forms of the injected plasmid. The levels of viral TK messenger RNA were insufficient for direct measurement by solution hybridization. However, message stability has been examined by injecting chicken ovalbumin or conalbumin mRNA into the cytoplasm of the growing oocyte and the fertilized egg. The stability of ovalbumin mRNA was 30 times greater and the stability of conalbumin mRNA was 120 times greater in the growing oocyte than in the fertilized egg [Ebert et al., 1985]. The rate of protein synthesis (as measured by the incorporation of radioactive amino acids into proteins) and the rate of decay of proteins were similar in the growing oocyte and fertilized egg (Table II). Thus, the 25–45-fold increase in TK activity obtained in fertilized eggs is most readily explained by greater transcriptional capacity. The difference in transcriptional activity might be much greater

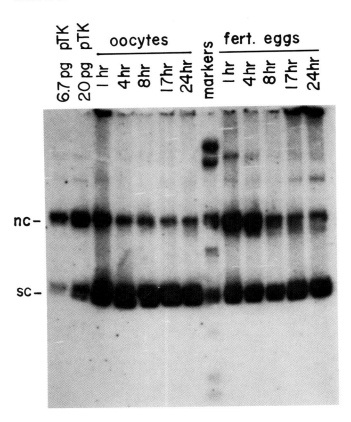

Fig. 1. Form of DNA following injection into the nucleus of mouse growing oocytes and fertilized eggs. Lanes 1 and 2 contain standards of HSV TK (7.9 kb); nc, nicked circles (or relaxed circles); sc, supercoiled DNA. Lane 3–7 contain DNA from growing oocytes that were harvested 1, 4, 8, 17, and 24 hr after injection into each nucleus of ~2 pl HSV TK DNA at 75 ng/μl. Each well received DNA from 48 to 66 oocytes. Based on the volume increase of the nucleus following injection and the number of oocytes, we estimate that a total of ~9.0 pg of DNA was injected into the nuclei. Lane 8 contains molecular weight markers (uncut λ DNA plus λ cut with HindIII). Lanes 9–13 contain DNA from fertilized eggs that were harvested 1, 4, 8, 18, and 24 hr after injection. Each well received DNA from 35 to 48 eggs; a total of ~6.4 pg HSV TK was injected into the pronuclei. The actual quantity of DNA recovered from the oocytes and eggs is greater than the amount injected into the nucleus, which probably represents leakage into the cytoplasm.

TABLE II. Synthesis and Stability of the Proteins in Growing Oocyte and Fertilized Egg

Stage of development	Incorporation of [^{3}H]amino acids [Ebert et al., 1985] (fmol/hr)	Protein half-life (hr)
Growing oocyte	17.5 ± 0.9	32.1 ± 3.1
Fertilized egg	28.1 ± 0.5	17.2 ± 1.9[a]

[a]Data for protein decay of fertilized egg from Merz et al. [1981]. Value for growing oocyte determined in the same manner.

than the difference in TK activity considering that mRNA is much more stable in the growing oocyte.

III. CHARACTERISTICS OF EXPRESSION OF GENES INJECTED INTO FERTILIZED MOUSE EGGS

Because it was not possible to measure mRNA production directly, TK enzyme activity was used to monitor the time course of transcription following gene injection. As is shown in Figure 2, viral enzyme activity does not appear until approximately 12 hr after the injection of the gene. Activity increases rapidly for the next 12 hr and then remains level during the following 24 hr. Data presented below suggest that transcription declines markedly with cleavage; thus the plateau in activity after 24 hr might reflect the balance between residual synthesis and degradation.

Fertilized ova collected on the first day of pregnancy (the day of the vaginal plug) are at the one-cell stage and will generally divide during the next 12–24 hr. However, following injection with DNA, cleavage fails to occur in some of the eggs, particularly when the concentration of DNA is high. Analysis of individual eggs demonstrated that there is not only variation in TK activity among eggs following gene injection but that there are consistently higher levels of activity in eggs that fail to cleave (data not shown). Based on the above results, we reasoned that maximum activity might be obtained if we could inhibit cleavage. Of several compounds examined, we found that 2 μM aphidicolin, an inhibitor of DNA polymerase [Ikegami et al., 1978; Aizawa et al., 1983], was the most effective. TK activity increased five- to tenfold when cleavage was blocked by aphidicolin, and normal regulation of the MK gene was not altered (Table III). Using this technique, and by including a reference gene, we have been able to assay a variety of metallothionein promoter modifications [Stuart et al., 1984].

Preliminary experiments indicated that when eggs were injected with a plasmid containing both the HSV TK and SV40 T-antigen gene, synthesis of

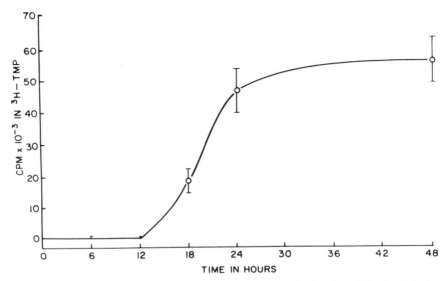

Fig. 2.   Increase in viral thymidine kinase activity in mouse fertilized ova following injection of a plasmid containing the TK gene. Approximately 140,000 plasmids were injected per nucleus. The eggs were then cultured in medium containing [³H]-thymidine and harvested at the specified times. Viral thymidine kinase activity was assayed as in Table I. Values are [³H]-TMP accumulated per egg; the points indicate the mean ± SEM for three experiments.

TABLE III. Effect of Cleavage and Aphidicolin on Thymidine Kinase Activity in Eggs

Stage of development	Aphidicolin[a] 2 μM in medium	Cadmium[b] 50 μM in medium	Formation of [³H]TMP (cpm × 10⁻³)
Two-cell	−	−	31 ± 4[c]
Two-cell	−	+	111 ± 23
One-cell	+	−	211 ± 52
One-cell	+	+	4,482 ± 1,381

[a]Aphidicolin treatment began immediately following collection and continued until the end of the culture period.
[b]Cadmium treatment began immediately following injection of 2,500 copies of pMK, a metallothionein-TK fusion gene [Stuart et al., 1984].
[c]Values for [³H]TPM are for 25 ova after 22 hr in culture and are the mean ± SEM.

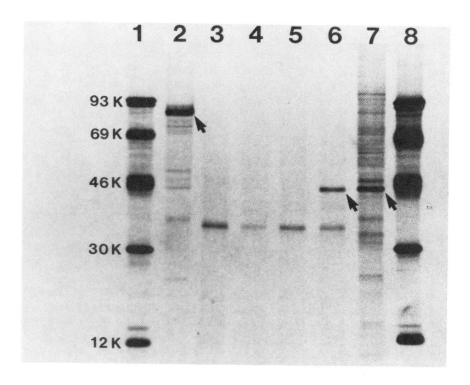

Fig. 3. Formation of HSV TK and SV40 T-antigen protein by fertilized mouse eggs following injection of a plasmid [Linnenbach et al., 1980] containing genes for both proteins. Eggs
were labeled with [^{35}S]-methionine, and extracts were made, precipitated with specific antibody, and separated by gel electrophoresis. A fluorograph of such a separation is shown. Gel
lanes are numbered on the top. Molecular weight (MW) indicated in thousands (K) on the left.
Lanes 1 and 8 are protein molecular weight markers. Lanes 2–4 are electrophoretic separations
of immunoprecipitates of SV40 T-antigen (MW 92 kd) from an SV40 transfected cell line, 35
control ova, and 35 plasmid injected ova, respectively. Lanes 5–7 are electrophoretic separations of immunoprecipitates of viral thymidine kinase (MW 45 kd) from 35 control ova, 35
plasmid injected ova, and a viral thymidine kinase transfected cell line, respectively. Location
of T-antigen and viral thymidine kinase proteins is indicated by arrows. The labeled bands at
33 kd in the lanes containing egg precipitates represent a slight contamination from the most
heavily labeled proteins in the egg [see Chen et al., 1980].

the TK protein could be detected (Fig. 3, lane 6) but SV40 T-antigen protein could not be detected (Fig. 3, lane 4). Because the SV40 T-antigen gene contains an intron, and the HSV TK gene does not, it seemed possible that the presence of introns could interfere with RNA processing in the eggs. Alternatively, the difference in promoters could alter transcription rates. Therefore, we compared TK genes with and without introns but with the same promoter. The mouse metallothionein I promoter/regulator was fused to either the HSV TK structural gene (no introns) or the chicken TK gene (six introns), and the ability of each gene to produce TK in the egg was determined. The chicken and HSV thymidine kinase appear to have similar activities [Merrill et al., 1984]. The data in Table IV indicate that the presence of introns is associated with decreased expression of the chicken TK gene.

IV. CONCLUSIONS

Several remarkable characteristics of early developmental stages of the mouse egg have emerged from these studies. The large increase in TK activity from genes injected into the fertilized egg above that found in prefertilized stages demonstrates one of the most significant differences yet identified between fertilized and prefertilized stages in the mammalian ovum. Although a variety of metabolic changes have been shown to occur associated with fertilization in sea urchin and frog embryos [Davidson, 1976; Gurdon, 1977], it has been difficult to identify clear-cut metabolic changes occurring in conjunction with mammalian fertilization. The release of calcium at the time of fertilization in mouse eggs is probably the best documented metabolic

TABLE IV. Expression of Metallothionein-Chicken TK and Metallothionein-HSV TK Genes Following Injection Into the Mouse Egg

Gene[a]	Introns	Cadmium 50 μM in medium	Formation of [^{3}H]TMP (cpm $\times 10^{-3}$)
MT cTK	+	−	247[b]
MT cTK	+	+	602
MT TK	−	−	1375
MT TK	−	+	5258

[a]Approximately 2,500 gene copies were injected into the male pronucleus of each egg. MT cTK is a fusion gene composed of the mouse metallothionein I promoter/regulator and the chicken TK structural gene. MT TK is a fusion gene composed of the mouse metallothionein I promoter/regulator and the HSV TK structural gene [Stuart et al., 1984].
[b][^{3}H]TMP formed from 25 eggs following 22hr incubation in culture medium containing [^{3}H]thymidine. Cadmium treatment began immediately following injection of the gene. Cadmium is a potent inducer of transcriptional activity for metallothionein fusion genes injected into eggs [Brinster et al., 1982].

change [Cuthbertson et al., 1981]. Although neither total RNA nor protein synthesis has been demonstrated to increase significantly following fertilization in the mouse [Brinster et al., 1976; Sherman, 1979], the rates of synthesis of several proteins have been shown to change soon after fertilization [Braude et al., 1979; Chen et al., 1980]. It is apparent from our data that the ability to transcribe an injected gene is much greater following fertilization. This increase in transcriptional ability is one of the first steps in activating the machinery necessary for transcription of new genes, which begins in the days following fertilization. The marked decrease in stability of injected mRNA in fertilized eggs [Ebert et al., 1985] also reflects an alternate mechanism for reprograming protein synthesis.

The data indicate there is little degradation of injected DNA during the initial 24 hr. The quantity of DNA present in the egg cells is significantly greater than calculated on the basis of nuclear volume change following injection. The excess DNA probably represents leakage into the cytoplasm. Much or all of this leakage occurs during and immediately following injection; the volume of the injected nucleus decreases significantly in the first 5–10 sec after injection. Therefore, the relaxed circle and supercoiled DNA is relatively stable in both the nucleus and cytoplasm. These finds are in contrast to those in the *Xenopus* oocyte, in which DNA injected into the cytoplasm is degraded rapidly [Gurdon, 1977; Gurdon and Melton, 1981].

Although there is little change in the form of the DNA during the 24 hr following injection, it is clear from Figure 2 that expression of the gene is not constant during this interval. Perhaps the initial lag period reflects assembly into chromatin, which is then followed by active transcription of the gene. Although chromatin assembly might account for low activity during the initial 12 hr, it is not clear why enzyme activity should plateau after 24 hr. Perhaps processes associated with subsequent cleavage are involved in gene inactivation; we have routinely observed lower TK activity in eggs that divide compared to those that do not. Furthermore, preventing DNA synthesis with aphidicolin allows a five- to tenfold increase in TK activity.

The comparison of SV40 TK and HSV-TK as well as chicken TK and pMK suggests that expression of genes with introns is impaired in fertilized eggs. However, we do not know whether this is owing to saturation of the enzymes involved in RNA splicing or reflects normal mechanisms involved in regulation of gene expression at these developmental stages.

Our studies employing the technique of gene microinjection into the mouse egg have revealed a dramatic increase in the ability of the egg cell to transcribe RNA polymerase type II genes following fertilization. This is clearly an important step in activation of the egg and is essential for the subsequent utilization of endogenous genes during development of the embryo.

ACKNOWLEDGMENTS

The plasmid containing the HSV TK and SV40 T antigen genes was generously provided by A. Linnenbach. We thank Paul Fallon for technical assistance and Carolyn Pope for typing the manuscript. We are grateful to our colleagues for helpful suggestions. Financial support was from the NIH and the NSF.

V. REFERENCES

Aizawa S, Loeb LA, Martin GM (1983): Mouse teratocarcinoma cells resistant to aphidicolin and arabinofuranosyl cytosine: Isolation and initial characterization. J Cell Physiol 115:9–14.

Bachvarova R (1985): Gene expression during oogenesis and oocyte development in mammals. In Browder L (ed): "Developmental Biology: A Comprehensive Synthesis, Vol. 1: Oogenesis." New York: Plenum Press, pp 453–524.

Braude P, Pelham H, Flach G, Lobatto R, (1979): Post-transcriptional control in the early mouse embryo. Nature 282:102–105.

Brinster RL (1973): Nutrition and metabolism of the ovum, zygote and blastocyst. In Greep RO (ed): "Handbook of Physiology, Section 7, Vol. II, part 2." American Physiological Society, Washington, pp 165–185.

Brinster RL, Chen HY, Trumbauer ME (1981a): Mouse oocytes transcribe injected Xenopus 5S RNA gene. Science 211:396–398.

Brinster RL, Chen HY, Trumbauer ME, Avarbock MR (1980): Translation of globin messenger RNA by the mouse ovum. Nature 283:499–501.

Brinster RL, Chen HY, Trumbauer ME, Paynton BV (1981b): Secretion of proteins by the fertilized mouse ovum. Exp Cell Res 134:291–296.

Brinster RL, Chen HY, Warren R, Sarthy A, Palmiter RD (1982): Regulation of methallothionein-thymidine kinase fusion plasmids injected into mouse eggs. Nature 296:39–42.

Brinster RL, Wiebold JL, Brunner S (1976): Protein metabolism in preimplanted mouse ova. Dev Biol 51:215–224.

Chen HY, Brinster RL, Merz EA (1980): Changes in protein synthesis following fertilization of the mouse ovum. J Exp Zool 212:355–360.

Cuthbertson KSR, Whittingham DG, Cobbold PH (1981): Free Ca^{++} increases in exponential phases during mouse oocyte activation. Nature 294:754–757.

Davidson EH (1976): "Gene Activity in Early Development, Second Ed." New York: Academic Press.

Ebert KM, Paynton BV, McKnight GS, Brinster, RL (1984): Translation and stability of ovalbumin messenger RNA injected into growing oocytes and fertilized ova of mice. J Embryol Exp Morphol 84:91–103.

Gurdon JB (1977): Egg cytoplasm and gene control in development. Proc R. Soc London series B 198:211–247.

Gurdon JB, Melton DA (1981): Gene transfer in amphibian eggs and oocytes. Annu Rev Genet 15:189–218.

Ikegami S, Taguchi T, Ohashi M, Oguro M, Nagano H, Mano Y (1978): Aphidicolin prevents mitotic cell division by interfering with the activity of DNA polymerase-alpha. Nature 275:458–460.

Linnenbach A, Huebner K, Croce CM (1980): DNA-transformed murine teratocarcinoma cells: Regulation of expression of simian virus 40 tumor antigen in stem versus differentiated cells. Proc Natl Acad Sci USA 77:4875–4879.

Merrill GF, Hauschka SD, McKnight SL (1984): tk enzyme expression in differentiating muscle cells is regulated through an internal segment of the cellular tk gene. Mol Cell Biol 4:1777–1784.

Merz EA, Brinster RL, Brunner S, Chen HY (1981): Protein degradation during preimplantation development of the mouse. J Reprod Fertil 61:415–418.

Sherman MI (1979): Developmental biochemistry of preimplantation mammalian embryos. Annu Rev Biochem 48:443–470.

Stuart GW, Searle PF, Chen HY, Brinster RL, Palmiter RD (1984): A twelve-base-pair DNA motif that is related several times in metallothionein gene promoters confers metal regulation to a heterologous gene. Proc Natl Acad Sci USA 81:7318–7322.

Index